T0201214

Special Publications 70

WOMEN IN THE GEOSCIENCES

Practical, Positive Practices Toward Parity

Edited by

Mary Anne Holmes
Suzanne OConnell
Kuheli Dutt

This Work is a co-publication between
the American Geophysical Union and John Wiley & Sons, Inc.

WILEY

This Work is a co-publication between the American Geophysical Union
and John Wiley & Sons, Inc.

Published under the aegis of the AGU Books Board

Brooks Hanson, Director of Publications
Robert van der Hilst, Chair, Publications Committee

© 2015 by the American Geophysical Union, 2000 Florida Avenue, N.W., Washington, D.C. 20009
For details about the American Geophysical Union, see www.agu.org.

Published by John Wiley & Sons, Inc., Hoboken, New Jersey
Published simultaneously in Canada

No part of this publication may be reproduced, stored in a retrieval system, or transmitted in any
form or by any means, electronic, mechanical, photocopying, recording, scanning, or otherwise,
except as permitted under Section 107 or 108 of the 1976 United States Copyright Act, without
either the prior written permission of the Publisher, or authorization through payment of the
appropriate per-copy fee to the Copyright Clearance Center, Inc., 222 Rosewood Drive,
Danvers, MA 01923, (978) 750-8400, fax (978) 750-4470, or on the web at www.copyright.com.
Requests to the Publisher for permission should be addressed to the Permissions Department,
John Wiley & Sons, Inc., 111 River Street, Hoboken, NJ 07030, (201) 748-6011, fax (201) 748-6008,
or online at http://www.wiley.com/go/permissions.

Limit of Liability/Disclaimer of Warranty: While the publisher and author have used their
best efforts in preparing this book, they make no representations or warranties with respect to
the accuracy or completeness of the contents of this book and specifically disclaim any implied
warranties of merchantability or fitness for a particular purpose. No warranty may be created or
extended by sales representatives or written sales materials. The advice and strategies contained
herein may not be suitable for your situation. You should consult with a professional where
appropriate. Neither the publisher nor author shall be liable for any loss of profit or any other
commercial damages, including but not limited to special, incidental, consequential, or
other damages.

For general information on our other products and services or for technical support, please contact
our Customer Care Department within the United States at (800) 762-2974, outside the United
States at (317) 572-3993 or fax (317) 572-4002.

Wiley also publishes its books in a variety of electronic formats. Some content that appears in print
may not be available in electronic formats. For more information about Wiley products, visit our
web site at www.wiley.com.

Library of Congress Cataloging-in-Publication data is available.

ISBN: 978-1-119-06785-6

Cover images:

*Troikken. A winters day holds a rare oportunity to view Mt. Fuji with complete clarity. Fuji-san normally
hides beneath clouds and snow filled winds but this day the great mountain decides to put on a show.
Titlezpix. Beautiful silhouette sunset at tropical sea.
Kali9. Close up of mature woman, 40s, working in industrial lab in chemical plant.
Lumenetumbra. The Wave at the Paria Wilderness in Northern Arizona.
Ericfoltz. The famous Great Smoky Mountains National Park, North Carolina.*

Printed in the United States of America

10   9   8   7   6   5   4   3   2   1

CONTENTS

Color plate section located between pages 88 and 89

CONTRIBUTORS

Ann E. Austin
Professor
Higher, Adult, and Lifelong Education
Michigan State University
East Lansing, Michigan

Linnea Avallone
Program Director
Division of Atmospheric and Geospace Sciences
National Science Foundation
Arlington, Virginia

Sarah Clem
MPOWIR Program Coordinator
Nicholas School of the Environment
Duke University
Durham, North Carolina

Kuheli Dutt
Assistant Director, Academic Affairs and Diversity
Lamont-Doherty Earth Observatory
Columbia University
Palisades, New York

Laura M. Edwards
Climate Field Specialist
South Dakota State University Extension
Aberdeen, South Dakota

Jennifer B. Glass
Assistant Professor
Georgia Institute of Technology
Atlanta, Georgia

A. Gannet Hallar
Associate Research Professor and Director
Storm Peak Laboratory
Division of Atmospheric Science
Desert Research Institute
Steamboat Springs, Colorado

Meredith G. Hastings
Assistant Professor
Department of Earth, Environmental and Planetary Sciences
Brown University
Providence, Rhode Island

Mary Anne Holmes
Professor of Practice
Department of Earth and Atmospheric Sciences
University of Nebraska–Lincoln
Lincoln, Nebraska

Rose Kontak
Program Coordinator
Environmental Change Initiative
Brown University
Providence, Rhode Island

Sandra L. Laursen
Co-Director and Senior Research Associate
Ethnography & Evaluation Research
University of Colorado Boulder
Boulder, Colorado

Susan Lozier
Professor, Physical Oceanography
Nicholas School of the Environment
Duke University
Durham, North Carolina

Dalinda Martinez
Graduate Student
Higher, Adult, and Lifelong Education
Michigan State University
East Lansing, Michigan

Suzanne OConnell
Professor
Department of Earth and Environmental Science
Wesleyan University
Middletown, Connecticut

Melissa Soto
Graduate Student
Higher, Adult, and Lifelong Education
Michigan State University
East Lansing, Michigan

Heather Thiry
Researcher
Ethnography & Evaluation Research
University of Colorado Boulder
Boulder, Colorado

Christine Wiedinmyer
Scientist III
National Center for Atmospheric Research
Boulder, Colorado

INTRODUCTION

Does Gender Parity Matter?

The geoscience workforce has a lower proportion of women in it (21%) compared to the general population of the United States (50%) and compared to the average of all other science (37%) or mathematics (26%) fields [*NSF*, 2011]. Our workforce is overwhelmingly white: 86% compared to 68% of the total U.S. population, one of the least diverse among all the other science, technology, engineering, and mathematics (STEM) fields. In short, the U.S. geoscience workforce lacks the rich diversity of our population. According to the 2011 U.S. Census Bureau, 88% of doctoral degrees in the geosciences are awarded to white students, with only about 5% awarded to students from underrepresented minority groups.

Does this low rate of diversity matter? Obviously, since you have opened this book and many geoscientists have contributed to it, many people think so and you likely think so, too. Feelings aside, what is the evidence that it really matters? Is it something we should put our limited time and resources into addressing, to have our workforce diversity more closely match the nation's? Does the geoscience enterprise really suffer if we never diversify to match population demographics?

The answer we contend is, of course, yes. Scott Page, an economist at the University of Michigan, uses mathematical modeling and case studies to show that diverse workplaces are more productive, more innovative, and more creative (2008). People with different backgrounds have different ways of looking at problems (what Page calls "tools"). In science, having more tools generates more working hypotheses, a necessary step in the scientific method. But not only do different types of people view problems differently, different types of people *ask different questions*, the fundamental first step in the scientific method. Bringing personal knowledge to the scientific endeavor means that different scientists sense (observe) differently, question differently, and hypothesize differently [*Selby*, 2006a,b].

Women in the Geosciences: Practical, Positive Practices Toward Parity, Special Publications 70.
First Edition. Edited by Mary Anne Holmes, Suzanne OConnell, and Kuheli Dutt.
© 2015 American Geophysical Union. Published 2015 by John Wiley & Sons, Inc.

Page [2007] points out that today, teams do more work (and science) rather than do lone individuals. Does diversity improve team performance? *Woolley and others* [2010] developed a measurement of group intelligence (termed c) and determined, surprisingly, that it does not correlate with either average individual intelligence of group members or with maximum individual intelligence (the "smartest person" in the group). Instead, they found that c significantly correlated with a measure of average social sensitivity of the group and negatively correlated with the presence of a few people in the group who dominated the conversation. The presence of women in the group increased the group's intelligence as measured by its ability to perform specific group tasks. These researchers hypothesized that in this study, the women's influence arose from their tendency to score higher on social sensitivity tests. Just having more people able to voice an opinion raised group intelligence.

Page [2007] found that when a team values diversity, a diverse work group improves the bottom line for corporations, perhaps as much as the actual ability of individual workers. In a diverse workforce, people's abilities are superadditive: if two people have different perspectives on a problem as well as different proposed solutions, the best solution may lie in a combination of the two solutions, an outcome not possible when only one brain works on the problem.

Govindarajan and Terwilliger [2012] found that a diverse team does the most effective research brainstorming. Like Page, they use the term *diversity* to include a range of expertise, ages, disciplines, and cultures.

Valian [2004] provides additional rationales for the benefits of gender parity in academia. Broadening the applicant pool for faculty positions maximizes the chances of hiring the best new faculty. The larger the pool, the greater will be the choice and the higher the likelihood of finding a well-qualified candidate.

Students benefit from a diverse faculty. Students who see someone on the faculty "like me," someone whose life they wish to emulate, are more likely to stay in the field. In addition, students benefit from working in diverse groups and with diverse faculty, as they will be working in a diverse workforce after graduation [*Valian*, 2004]. The benefits of being a scientist are great: scientists earn more than nonscientists and are more likely to be employed. And as scientists, we know the joy of doing science that no other field of endeavor provides.

Diversity of the geoscience workforce matters because we need a variety of minds asking a variety of questions and posing a variety of solutions. Diversity of the geoscience workforce matters because the U.S. population continues to diversify: nonwhite children became the majority of one-year-olds in 2010. We need to attract new majors and new geoscientists from the population that exists today and tomorrow or we will find our classrooms and consequently the geoscience workforce shrinking.

Paying attention to the factors that promote gender equity in departments improves the workplace for *all* faculty [*Valian*, 2004]. When we discover that mentoring, advocacy, and power networks omit women and people of color, and we construct mentoring programs for early and mid-career faculty, these benefit *all* faculty. When we address dual-career issues for women, we address dual-career issues for men, too. More than half of STEM men (56%) are married to a STEM woman [*Schiebinger*, 2008]. As more women and people of color have received PhDs and expect an inclusive workplace, the majority's perception of what makes a good work environment has evolved, too. We are not the same academy that we were 10 years ago.

Professional science societies recognize the value of diversity. For example, The American Association for the Advancement of Science has issued a statement with the Association of American Universities in support of diversity-enhancing programs (http://php.aaas.org/programs/centers/capacity/documents/Berdahl_Essay); the American Geophysical Union has a Diversity Plan (http://education.agu.org/diversity-programs/agu-diversity-plan/); the Geological Society of America adopted a position statement to embrace a diverse workforce (http://www.geosociety.org/positions/pos15_Diversity.pdf), and the American Association of Petroleum Geologists has held panels on making the bottom-line case for diversity in the petroleum industry (http://www.aapg.org/explorer/2010/06jun/regsec0610.cfm).

Yet despite broad support for the concept of gender parity, there has been little actual change in the demographics of the geoscience faculty (see chapters 1 and 2).

Why Are the Geosciences Lagging in Gender Parity?

We show from the literature through the rest of this volume that lack of gender parity is not unique to the geosciences and that, for all STEM fields, gender parity is *not* a "pipeline issue": simply adding more women to one end of the pipeline, such as PhD recipients, has not effected meaningful change in the numbers of women on the STEM faculty. Nor is the answer simply "women prefer to have families," as the numbers of single women or women with no children are not increasing on the faculty. Policies and procedures of academic institutions, as well as how we perceive and interact with each other, play important roles in whether we can achieve parity. The academy is set up for an "ideal worker" who is currently in the majority [*Williams*, 2000]. Our selection processes, those that determine who gets encouraged to enter graduate school, to complete the PhD and postdoc, and to win the job, contribute to the leaky pipeline [*Georgi*, 1999]. The academy needs to change to accommodate a variety of types of workers.

Chilly climates continue to contribute to women's attrition from the geosciences. By "climate" we mean the factors in the workplace that enable us to find meaning and joy in our work. It is an important component of job satisfaction. A variety of factors can contribute to chilly climates for women. The literature is replete with examples of women's accomplishments being discounted and ignored [*Lincoln et al.*, 2011; and see, in references, the AWIS AWARDS project to increase the number of women nominees for national awards]. In addition, women are more likely to serve on committees that are perceived as nurturing (e.g., undergraduate advisor) as opposed to committees that wield influence on academic processes, such as promotion and tenure and graduate committees [e.g., *Misra et al.*, 2011]. Women typically have higher service loads and take these on at earlier stages in their career [*Misra et al.*, 2011], in part because they are asked to serve as the "diversity" component on every committee. Women tend to be interrupted more at meetings, tend to have lower salaries, sometimes as a result of their not negotiating sufficiently [*Bilimoria and Liang*, 2011; *Valian*, 2005]. As *Valian* [2005] puts it, "Each example [of chilly climate] . . . is a small thing. One might be tempted to dismiss concern about such imbalances as making a mountain out of a molehill. But mountains *are* molehills, piled one on top of another over time."

Student Perspective

As we wrote this volume, younger women provided us with plenty of examples of the chilly climate they experience. Below, a few examples:

"The professor told the class that women really weren't that good for geology because they value family more than anything else. The only person who objected was a male postdoc who said he thought family was just as important to men."

"The male presenters frequently made good-natured and humorous comments about other male lecturers that were present in the lecture hall. They used each other's first names. The one time I heard a male lecturer make a comment about a female lecturer that was present, he did not use her name but referred to her as 'that woman.'"

"A lecture given by a woman was interrupted by male organizers announcing the arrival of a new (male) lecturer and the departure of another (male) lecturer. Later on the same talk was again interrupted by another departing (male) lecturer wanting to announce he was leaving. No talk given by a man was interrupted by such departures and arrivals."

"The female participants of the summer school were sometimes referred to as 'girls.' Male participants were not addressed as 'boys' or 'guys,' at least never within my hearing."

"During an evening event, a medal was given to a distinguished male scientist. . . . After the talk the organizers took photos of the medal-winning scientist. They addressed the audience and asked for 'girls' to step up and have their photo taken with the awardee."

"While I was completing an assignment in an all-female group, one of the male lecturers stopped by to inquire how we were doing, and then made a loud public comment about the beauty of our group. I heard no such comments about the appearance of the male participants."

"In three different talks, the lecturers had included in their overheads a photo of a woman in revealing clothing. In all cases, the woman had a 'conventionally beautiful' body type and general appearance. I saw only one photo that depicted a man in sparse clothing, and in that case the man was very obese. I got the feeling that female bodies were shown not only to illustrate a point, but also because they were thought to be pretty to look at (and amusing in a scientific context). The man's photo was also there to make a humorous point, but in his case the humor largely stemmed from the fact that he was very fat (and very fat guys are supposedly funny)."

The signature file from the e-mail of a (male) chair of an earth sciences department:

The primary duty of the University to a student is to provide him with such instructors as will make him realise that the responsibility for progress is his own and no one else's.

S.E. Whitnall, 1933

"This phrase. This 'you will ruin your career if.' It's false. It's a total, complete lie. And it really upsets me to watch so many young, promising scientists agonize and fall prey to it. Because the correct phrase is not 'you will ruin your career if,' the correct phrase is 'your career (in a TT position at an R1 institution) will be a lot easier if.'"

Isn't This Issue Behind Us by Now?

The above examples provided by women students are fairly convincing that we have not yet fully thawed the chilly climate for women. In addition, Nancy Hopkins, the author of the now-famous "MIT Study" that brought gender inequity on that campus to light [*Hopkins*, 1999; *Hopkins*, 2007], demonstrated that when there was agitation for adding women to the faculty, excellent women were found and hired at MIT (Figure 0.1). When the agitation waned, hiring leveled

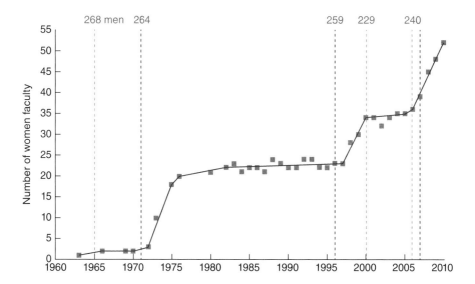

Figure 0.1 Women faculty in the School of Science at MIT (1960–2010). The numbers of women increase only when effort is focused on their recruitment and retention. Between 1970 and 2010, the percentage of women faculty at MIT increased from 8% to 19% [*from Conrad et al.*, 2011]. For color detail, please see color plate section.

off. A renewal of agitation increased hiring again. Looking at women's composition on the geoscience faculty in the earlier part of this decade, we noticed that many departments had one woman on the faculty at approximately midcareer. We all sort of rushed to hire our woman and then neglected the issue from then on. The ADVANCE program at NSF (see Chapter 4) has renewed the agitation to pay attention to this issue. When we stop paying attention, we make no progress.

The Pipeline Metaphor

Many women object to the concept of a pipeline: that we input students at one end and some proportion emerges ready for faculty positions. They do not wish to be considered passive in the motion from one end to the other, and particularly do not wish to be considered passive drops of water that leak out of the system.

A better metaphor for the process of developing new scientists is the interstate highway system: there are many ways to enter the science enterprise, beginning with a community college or beginning with entry into a Research I institution. The various ways to enter the path are "on-ramps." There are various stages at which a student might exit (off-ramps) and perhaps reenter via a different on-ramp at another time. Students might take "rest stops" via working in private industry or staying at home to start a family. Interstates lead to multiple destinations: academia is not the only endpoint for geoscience students. Exiting, entering, resting, and reaching a destination all imply some agency on the part of the participant.

Not all interstates are the same; some are state of the art with clear signposts and directions; others are in need of repair, perhaps rerouting, better on-ramps, or at least, better signage.

Contents of This Volume

The remainder of this volume will discuss research-based reasons for the lack of gender parity and research-based strategies to achieve gender parity. In Section I we look at data on gender parity in the geoscience student body and faculty. Chapter 1 looks at the gender composition of the recipients of geoscience degrees. Chapter 2 looks at the statistics of female faculty in Carnegie top-tier geoscience departments across the U.S.

Section II provides a conceptual framework for understanding and addressing gender parity issues. Specifically, chapter 3 explores Risman's theory of gender as a social structure that allows us to categorize types of barriers to women's entry, retention, and advancement in the geosciences.

Section III looks at various lessons learned from NSF-funded ADVANCE programs across the U.S. and the best practices learned from these programs, and summarizes the experiences of various institutions' progress made towards gender parity. This section first provides an overview of the NSF ADVANCE program, followed by examples of institutional, individual, and interactional strategies.

Chapter 4 provides an overview of NSF's ADVANCE program. Chapter 5 summarizes work done at ADVANCE institutions, that is, those that received an ADVANCE Institutional Transformation award. This chapter focuses on the effectiveness and long-term viability of organizational change efforts to create institutional environments that are conducive to the success of women as well as men in STEM. Chapter 6 presents an overview of the successful institutional transformation process of Columbia University's Lamont-Doherty Earth Observatory. Chapter 7 looks at how faculty appointments can be made more flexible and therefore conducive to retaining women; specific examples include stop-the-clock provisions, the option to work part-time, and dual-career appointments. Chapter 8 looks at the provision of on-campus lactation facilities and access to day care; since women bear a disproportionately higher burden of familial responsibilities, such facilities will help to retain them in STEM.

Chapter 9 discusses implicit bias, stereotype threat, imposter syndrome, and how these affect efforts to diversify the workforce. Chapter 10 looks at the best practices for recruiting diverse faculty by diversifying the applicant pool. Chapters 11 through 13 focus on mentoring. Chapter 11 discusses multiple and sequential mentoring, while chapters 12 and 13 expand upon intensive mentoring programs: ASCENT (Atmospheric Science Collaborations and Enriching Networks) and MPOWIR (Mentoring Physical Oceanography Women to Increase Retention). These two programs serve as excellent models not just for mentoring but also on how to increase transparency of processes in academia that lead to success of new faculty. Chapter 14 explains the Earth Science Women's Network, ESWN, a peer-mentoring network for women geoscientists particularly targeting early-career women.

Some of what we write in this volume also applies to the issue of race and ethnicity parity in the geosciences' workforce. We focus on gender parity for this volume because it is time, after more than a decade of focused research through the ADVANCE program, to pull together a what-, why-, and how-to-proceed handbook. So far, no similar body of work exists to address racial and ethnicity underrepresentation. We hope that you find this volume useful and we welcome any constructive feedback.

ACKNOWLEDGEMENTS

We wish to thank all of our colleagues who contributed to this volume, to the reviewers of each contribution, and to all of our colleagues who discussed and debated these issues with us, and thanks to our colleagues at AGU who helped see this book through to publication. We wish to acknowledge the financial support of the National Science Foundation through NSF ADVANCE Grants #0620101 and 0620087.

REFERENCES

AWIS AWARDS Project: http://www.awis.org/displaycommon.cfm?an=1&subarticlenbr= 397. Last accessed January 2014.

Bilimoria, D., & Liang, X. (2011). Gender Equity in Science and Engineering: Advancing Change in Higher Education. Routledge Studies in Management, Organizations and Society. Routledge, Taylor & Francis Group.

Conrad, J., N. Hopkins, T. Orr-Weaver, M. Potter, P. Rizzoli, H. Sive, G. Staffilani, and J. Stubbe (2011), Massachusetts Institute of Technology (MIT) Report on the Status of Women Faculty in Science and Engineering, 2011, http://web.mit.edu/newsoffice/images/documents/women-report-2011.pdf. Last accessed January, 2014.

Georgi, H. (1999), A tentative theory of unconscious discrimination against women in science. In *Who Will Do the Science of the Future? A Symposium on Careers of Women in Science*, pp. 45–48. NAS Committee on Women in Science and Engineering, National Research Council. National Academy Press, Washington, DC.

Govindarajan, V., and J. Terwilliger (2012), Yes, You Can Brainstorm Without Groupthink. Harvard Business Review blog network. http://blogs.hbr.org/cs/2012/07/yes_you_can_brainstorm_without.html. Accessed 26 July 2012.

Hopkins, N. (1999), A study on the status of women faculty in science at MIT, *MIT Faculty Newsletter*, *11*(4), March.

Hopkins, N. (2007), Diversification of a university faculty: Women faculty in the MIT schools of science and engineering, *New England Journal of Public Policy*, *22*(1), 11.

Lincoln, A. E., S. H. Pincus, and P. S. Leboy (2011), Scholars' awards go mainly to men, *Nature*, *469*(7331), 472–472.

Misra, J., J. H. Lundquist, E. Holmes, and S. Agiomavritis (2011), The ivory ceiling of service work, *Academe*, *97*(1), 22–26.

National Science Foundation, National Center for Science and Engineering Statistics (2011), *Science and Engineering Degrees: 1966–2008*. Detailed Statistical Tables NSF 11-316. Arlington, VA. Available at http://www.nsf.gov/statistics/nsf11316/.

Page, S. (2007), Making the difference: Applying a logic of diversity, *Academy of Management Perspectives*, November, 6–20.

Page, S. (2008), *The Difference: How the Power of Diversity Helps Create Better Groups, Firms, Schools, and Societies*, Princeton University Press, Princeton, New Jersey.

Schiebinger, L., A. D. Henderson, and S. K. Gilmartin (2008), Dual-Career Academic Couples: What Universities Need to Know. Clayman Institute for Gender Research, Stanford University. http://www.stanford.edu/group/gender/Publications/index.html.

Selby, C. C. (2006a), The Missing Person in Science: Inquiry Begins with I, *New York Academy of Sciences Update* 10–13. http://cecilyselby.blogspot.com/2012/02/missing-person-in-science-inquiry.html

Selby, C. C. (2006b), What Makes It Science: A Modern Look at Scientific Inquiry, *Journal of College Science Teaching*, *35*, 8–11. http://www.nsta.org/store/product_detail.aspx?id=10.2505/4/jcst06_035_07_8. http://cecilyselby.blogspot.com/2012/02/ what-makes-it-science-modern-look-at.html.

Valian, V. (2004), Beyond gender schemas: Improving the advancement of women in academia, *Natl Women's Studies Assoc Journal*, *16*(1), 207–220.

Valian, V. (2005). Beyond gender schemas: Improving the advancement of women in academia, *Hypatia*, *20*(3), 198–213.

Williams, J. (2000). *Unbending Gender: Why Family and Work Conflict and What to Do About It*. Oxford University Press, New York.

Woolley, A. W., C. F. Chabris, A. Pentland, N. Hashmi, and T. W. Malone (2010), Evidence for a collective intelligence factor in the performance of human groups, *Science*, *330*, 686–688. doi: 10.1126/science.1193147

SECTION I: THE DATA

1

WHO RECEIVES A GEOSCIENCE DEGREE?

Mary Anne Holmes

Department of Earth and Atmospheric Sciences, University of Nebraska–Lincoln, Lincoln, Nebraska

ABSTRACT

To match applicant pools for faculty positions, and ultimately, faculties with the available pool, the student population, we need data on who gets a geoscience degree. The National Science Foundation (NSF) provides these data; they reveal that in the past 10 years, 35–40% of geosciences bachelor's and doctoral degrees are awarded to women; yet, less than 30% of geoscience assistant professors at doctoral-granting institutions are women. The principal leak in the academic pipeline, then, occurs at the entry-level hiring stage.

How many women should be on geoscience faculty? We propose that the proportion of women on the geoscience faculty should approximate the proportion who earn geoscience degrees. An analysis of NSF data on gender and race/ethnicity of STEM degree recipients in the U.S. in the last 10 years reveals that 35% to 40% of geosciences bachelor's and doctoral degrees were awarded to women. Yet less than 30% of geoscience assistant professors at doctoral-granting institutions are women.

1.1. Bachelor's Degrees

The National Science Foundation and the American Geosciences Institute collect data on who receives what degree in STEM and earth and atmospheric sciences (EAS) fields, respectively (http://www.nsf.gov/statistics/sestat/; http://www.agiweb.org/workforce/). NSF's data extend from 1967 to the present (no data were supplied for 1999). Undergraduate degrees awarded to women in

Women in the Geosciences: Practical, Positive Practices Toward Parity, Special Publications 70. First Edition. Edited by Mary Anne Holmes, Suzanne OConnell, and Kuheli Dutt.
© 2015 American Geophysical Union. Published 2015 by John Wiley & Sons, Inc.

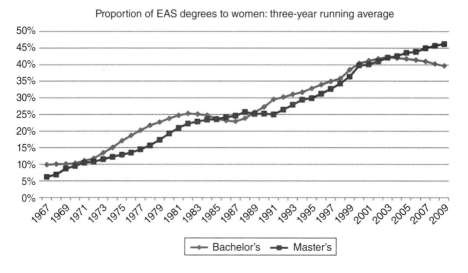

Figure 1.1 Proportion of Bachelor's and Master's degrees in EAS awarded to women. Data from *NSF*, 2013. For color detail, please see color plate section.

EAS fluctuate from 1967 to the present, but there is an overall upward trajectory, from 10% in 1967 to around 40% in 2010 (most recent data available; [*NSF*, 2011, 2013]) (Figure 1.1). Fluctuations appear to coincide with perceptions of the job market; that is, when the "oil bust" occurred in the mid-1980s, enrollments in geoscience programs declined rapidly. The decline was steeper for women than for men as indicated by the decrease in percentage of bachelor's degrees awarded to women during the oil bust (Figure 1.1). We have no explanation for why women would disproportionately not choose or would leave EAS when the oil job market declined. No studies of this phenomenon exist to date.

With time, the downward trend of the mid-1980s reversed. However, the proportion of women receiving EAS bachelor's degrees reversed again from its peak of 43% in 2002 (Figure 1.1). We know of no data that explain the current decline. In general, EAS underrecruits women to the field: since 1981, more than 50% of earned bachelor's degrees have been awarded to women; since 2000, more than 50% of earned STEM bachelor's degrees have been awarded to women [*NSF*, 2013]. The higher percentages are in the life sciences; the physical sciences and engineering continue to underrecruit women even more than does EAS.

Why would women not be attracted to EAS as a major? We asked focus groups of students for their ideas on this question, and both men and women cited their appearance, their clothing, as a turn-off to some portion of the student body. "We wear Carhartts and hiking boots and don't wear makeup" were the sorts of comments the students made. They told anecdotes of their roommates in other disciplines noticing our appearance and sometimes making disparaging or humorous remarks of the "field look." This phenomenon deserves more and better study; we suspect that there are additional explanations for the underrecruitment that our research did not reveal. Positive things we can do to increase recruitment of the underrepresented are to focus on "critical incidents" in the geosciences pipeline, as detailed in *Levine and others* [2007].

1.2. Graduate Degrees

The proportion of women who receive a master's degree in EAS closely tracks the proportion receiving a bachelor's until the mid-2000s, when a greater proportion of women receive master's than bachelor's degrees in EAS (Figure 1.1). These data demonstrate that until the mid-2000s, EAS did a great job of equably recruiting students by gender from bachelor's programs into master's programs. This is not true of the physical or biological sciences: both disciplines lose women from their pipelines after the bachelor's degree [*NSF*, 2013; see chap. 1 for discussion of using the pipeline metaphor]. Why men are now being disproportionately lost from bachelor's to master's programs needs further study. Unless they are heading straight to PhD programs, this trend is cause for concern.

The proportion of women who receive a PhD in EAS declines from the proportion who receive a bachelor's or master's degree (Figure 1.2). As for most STEM disciplines, women leak from the pipeline disproportionately between the bachelor's and the PhD. When asked to explain this decline, geoscientists in focus groups in 2002 provided gendered responses: men mentioned "societal pressures"

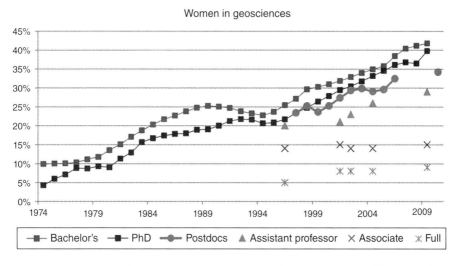

Figure 1.2 Proportion of women at various stages in the geoscience workforce pipeline. Student and post-doc data from *NSF*, 2013. Bachelor's degrees are forwarded by seven years to compare with PhD recipients. Faculty data from *AGI*, 1996–2012, for PhD-granting institutions. Bachelor's and Master's granting institutions have 3–5 higher percentage points of women faculty than doctoral granting institutions. For color detail, please see color plate section.

on women to have families, while women cited chilly department climates and the lack of structural support, such as daycare facilities [*Holmes et al.,* 2008].

1.3. On to the Profession

The proportion of women in postdoctoral positions closely matches the proportion who receive a PhD, indicating no or a small loss in the pipeline between PhD and postdoc [Figure 1.2; *NSF*, 2013].

The greatest leak (off-ramping) leading to academic positions occurs at hiring women into assistant professor positions (Figure 1.2). Research demonstrates that women feel both "push" factors for leaving the field between PhD and faculty position, and "pull" factors. "Push" factors are external factors such as implicit bias (see chap. 3), pressure from family or society to leave, lack of mentorship and encouragement to proceed in her career, lack of structural support for child care, and immobility due to partner's position, to name a few of these factors. "Pull" factors are those in her own life, personal circumstances that preclude her ability, interest, or desire to stay on an academic track. These might be a desire to care for family members (elders, siblings, or other family) or the overwhelming sense of a need to focus attention on a newborn.

Based on these data, applicant pools for faculty positions, and ultimately, the faculty itself, should have around 30% to 40% women in them to match the supply produced at the PhD and postdoc levels. We suggest strategies to increase the diversity of applicant pools in chapter 10.

The next chapter analyzes the faculty of the top 100 geoscience graduate programs in the U.S. as a sort of scorecard to see how we are progressing in creating a faculty that looks like our student body.

REFERENCES

American Geosciences Institute, *Directory of Geoscience Departments.* Issues from 1997, 2001, 2002, 2005, 2010, 2011, and 2012.
Holmes, M.A., S. OConnell, C. Frey, and L. Ongley (2008), Gender imbalance in U.S. geoscience academia, *Nature Geoscience,* *1*(2), 79–82. doi:10.1038/ngeo113.
Levine, Roger, S. Gonzalez, S. Cole. M. Fuhrman, and K. C. LeFlock (2007) The geoscience pipeline: A conceptual framework, *Journal of Geoscience Education,* *55,* 458–468.
National Science Foundation, National Center for Science and Engineering Statistics (2011), *Science and Engineering Degrees: 1966–2008.* Detailed Statistical Tables NSF 11-316, Arlington, VA. Available at http://www.nsf.gov/statistics/nsf11316/.
National Science Foundation, National Center for Science and Engineering Statistics (2013), *Women, Minorities, and Persons with Disabilities in Science and Engineering: 2013.* Special Report NSF 13-304, Arlington, VA. Available at http://www.nsf.gov/statistics/wmpd/.

2

WE ARE THE 20%: UPDATED STATISTICS ON FEMALE FACULTY IN EARTH SCIENCES IN THE U.S.

Jennifer B. Glass

Georgia Institute of Technology, Atlanta, Georgia

ABSTRACT

This paper presents data on the numbers of female and male professors at the 106 top US earth science PhD-ranting graduate programs during the 2010–2011 academic year. Overall, 20% of earth science faculty at PhD-granting research universities were women (470 female faculty members out of 2,324 total). By rank, 36% of assistant professors, 24% of associate professors, and 13% of full professors were women. Large ranges in percentages (0%–40%) of female professors were observed between departments. No geographic trends were observed, nor was there any correlation between the national ranking of department and the percentage of women faculty. A small positive correlation between the size of the department and the percentage of female faculty was present as department sizes increased from 5 to 30 faculty members, and a small decline occurred between 30 to 50 faculty. Percentages of tenured female faculty were generally lower than the total percentage of female faculty members in each department. The top 5 departments in terms of percentages of female faculty were SUNY Buffalo Department of Geology (40%), Louisiana State University–Baton Rouge Department of Geology and Geophysics (40%), University of New Hampshire Department of Earth Sciences (37%), University of Massachusetts–Amherst Department of Geosciences (36%), and University of Nevada–Las Vegas Department of Geoscience (35%).

Women in the Geosciences: Practical, Positive Practices Toward Parity, Special Publications 70. First Edition. Edited by Mary Anne Holmes, Suzanne OConnell, and Kuheli Dutt.
© 2015 American Geophysical Union. Published 2015 by John Wiley & Sons, Inc.

"I have always claimed that there was no merit in being the only one of a kind."

—**Florence Bascom** (1862–1945), first woman PhD from Johns Hopkins University, first woman geologist hired by U.S. Geological Survey, first woman officer of Geological Society of America, Bryn Mawr professor, founder of the Bryn Mawr geology department and mentor to numerous prominent female geologists. She modeled the geology program at Bryn Mawr on programs at male colleges, and insisted that her female students conduct field work, despite the fact that women's participation in geology had previously been primarily indoors (paleontology, cartography, etc). Many of the women geologists in the first part of the 20th century were followers of Florence Bascom [*Clary and Wandersee*, 2007].

Earth science is of vital importance to society: geoscientists strive to predict earthquakes and volcanic eruptions, forecast the effects of global climate change, and understand the evolution of life and global biogeochemical cycles through time, just to list a few research themes. Many women are fascinated by these topics, as illustrated by the fact that nearly 50% of the bachelor's degrees in earth science departments are granted to women [*Holmes et al.*, 2008].

The percentage of female graduate students in earth sciences is also relatively high: around 40% in the 2000s. The story changes going from graduate school to postdoctoral programs, and especially from postdoctoral programs into assistant professorships. Research has shown a large leak in the pipeline in between graduate school and assistant professorships [*Holmes and O'Connell*, 2003; *Holmes et al.*, 2008], implying that academic earth science is a less attractive career choice for female PhDs. This gap is not filling at nearly the rate expected if the problem was simply the lag time needed for the increased numbers of female PhDs to climb the ranks in academia [*Holmes et al.*, 2008].

I compiled data on the numbers of female and male professors at all ranks (assistant, associate, and full, as well as research professors at the same three ranks) for the 106 top U.S. earth science PhD-granting graduate programs from the 2011 *U.S. News & World Report* college rankings, with a minimum of 5 faculty in the department and a maximum of 50 faculty. These data were obtained by counting the numbers of female and male professors listed on the faculty pages of each department's webpage; counts were made between November 2010 and May 2011. Adjunct and emeritus professors were not counted. Taken all together, 20% of earth science faculty at PhD-granting research universities were women (470 female faculty members out of 2,324 total); by rank, this varied from 36% for assistant professors (33% for assistant research professors), 24% for associate professors (30% for associate research professors), and 13% for full professors (10% for full research professors) (Figure 2.1). These numbers are up ~10% across the ranks from a 2002–2003 dataset, which found that on average 12% of the total earth science faculty were female: 26% female assistant professors, 14% female associate professors, and 8% female full professors [*de Wet et al.*, 2002; *Holmes*

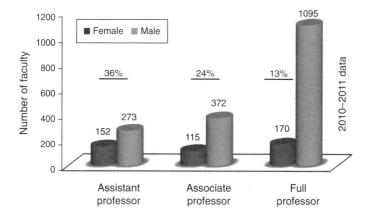

Figure 2.1 Numbers of female and male faculty members by rank at the 106 top-ranked PhD-granting geoscience departments. Data for 2010–2011 academic year. For color detail, please see color plate section.

and OConnell, 2003; *Holmes et al.*, 2008]. However, the 2010–2011 data remain well under the ultimate goal of 50% female earth science faculty at all ranks.

On a departmental level, there was a large range in the percentages of total faculty who were women, from 0% to 40% (Figure 2.2). One might suppose that departments with the most women are concentrated in a certain portion of the country. In fact, there are no clear geographic trends in the percentages of female faculty (Figure 2.3). There is also no correlation between ranking of department and the percentage of women faculty: the two top-ranked earth science graduate programs at Caltech and MIT have 22% and 18% female faculty, respectively, whereas two of the lowest ranked programs at University of Alabama and Baylor University have 27% and 7% female faculty, respectively. There was a very loose positive correlation between the size of the department and the percentage of female faculty as department sizes increased from 5 to 30 faculty members, and then a small decline in the percentage when the department size increased between 30 and 50 faculty. Percentages of tenured female faculty (associate and full professor) are generally lower than the total percentage of female faculty members in each department, with a few notable exceptions (Colorado State University: 50%, where 50% of the tenured professors are women; U. Nevada–Las Vegas: 44%; Georgia Tech, 38%; University of Wyoming: 27%; University of Wisconsin: 24%; Figure 2.2). This is important because women may be attracted to departments where there are already a significant number of senior female faculty members.

The top five departments in terms of percentages of female faculty in 2010–2011 were SUNY Buffalo Department of Geology (40%), Louisiana State University–Baton Rouge Department of Geology and Geophysics (40%), University of New Hampshire Department of Earth Sciences (37%), University of Massachusetts–Amherst Department of Geosciences (36%), and University of Nevada–Las Vegas

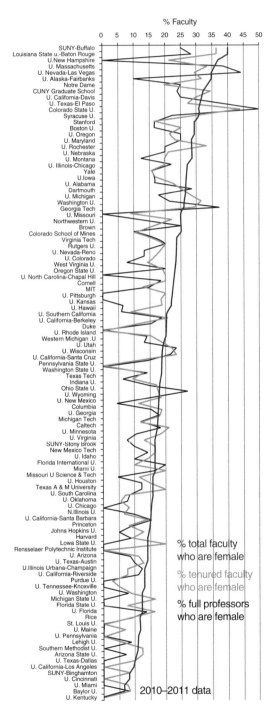

Figure 2.2 Percentage of female faculty by institution for 106 top-ranking PhD-granting geoscience departments in the U.S. Data for 2010–2011 academic year. For color detail, please see color plate section.

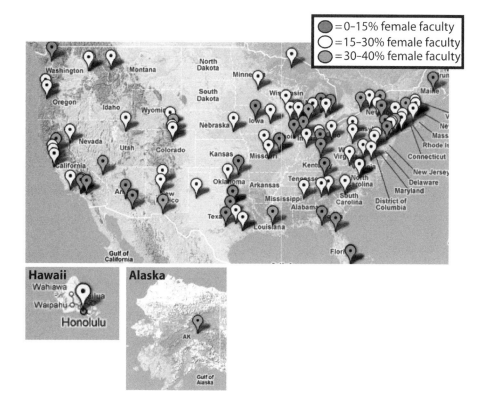

Figure 2.3 Geographic location of top-ranked geoscience departments in the U.S. labeled by color based on percentage of total faculty who are women. For color detail, please see color plate section.

Department of Geoscience (35%). The bottom five departments were SUNY Binghamton Department of Geological Sciences (8%), University of Cincinnati Department of Geology (8%), University of Miami Department of Geological Sciences (7%), Baylor University Department of Geology (7%), and University of Kentucky Department of Earth and Environmental Sciences (0%).

This chapter is reprinted from a previous publication: Glass, J. B. (2011), Increasing the recruitment and retention of women in academic geosciences: Where we are and where we should be, Association for Women in Science, 42, *24–27.*

REFERENCES

Clary, R.M. and J. H. Wandersee (2007), Great expectations: Florence Bascom (1842–1945) and the education of early US women geologists, in C. V. Burek and B. Higgs (eds.), *The Role of Women in the History of Geology*, Geological Society, London, Special Publications, *281*, 123–135.

de Wet, C. B. G. M. Ashley, and D. P. Kegel (2002), Supplement to Nov. 2002 *GSA Today*
12, 1–7, http://www.geosociety.org/pubs/gsatoday/0211clocks/0211clocks.htm.

Holmes, M.A., and S. OConnell (2003), Where are the women geoscience professors?
Report on the NSF/AWG Foundation Sponsored Workshop, September 25–27, 2003.

Holmes, M.A., S. OConnell, C. Frey, and L. Ongley (2008), Gender imbalance in US
geoscience academia, *Nature Geoscience*, *1*, 79–82.

U.S. News & World Report (2011), How U.S. News calculates its best colleges rankings,
http://www.usnews.com/education/best-colleges/articles/2012/09 /11/how-us-
news99calculates-its-best-colleges- rankings.

SECTION II: A FRAMEWORK TO ADDRESS THE ISSUE

3

A SOCIOLOGICAL FRAMEWORK TO ADDRESS GENDER PARITY

Mary Anne Holmes

Department of Earth and Atmospheric Sciences, University of Nebraska–Lincoln, Lincoln, Nebraska

ABSTRACT

In this chapter, we offer a sociological framework within which to test ideas and suggestions for why the leak in the pipeline is occurring. We explore Risman's theory of gender as a social structure, which categorizes barriers to participation into three groups: individual, interactional, and institutional. Barriers classified into these three categories can overlap, necessitating multiple responses to address the leak in the academic pipeline.

The data presented in the previous two chapters demonstrate that women are disproportionately lost from academic geoscience between the PhD and the first academic job, the assistant professor. This gap has persisted for at least the last decade; waiting for the pipeline to passively supply women into faculty positions has clearly not worked to date. What would stem the loss? The problem is a complex one: if there were one explanation for women's loss from the academy, geoscientists would have fixed it by now. What, then, are the reasons why this loss of women might be occurring?

In this chapter we offer a sociological framework within which to test ideas and suggestions for why the loss is occurring. In addition, it can help us focus attention in the right places and generate strategic programs to increase the number of women retained in geoscience.

3.1. Introduction

Ask any geoscientist why there are so few women on the faculty, and you will hear a variety of reasons: women choose to not go into academia (they either choose the higher salaries of industry or choose to have families); women

Women in the Geosciences: Practical, Positive Practices Toward Parity, Special Publications 70. First Edition. Edited by Mary Anne Holmes, Suzanne OConnell, and Kuheli Dutt.
© 2015 American Geophysical Union. Published 2015 by John Wiley & Sons, Inc.

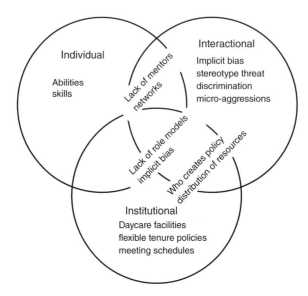

Figure 3.1 Individual, interactional, and institutional barriers to gender parity may overlap.

see few role models; women lack the networks and mentors that men have; women are absent from the highest math achievers; we're waiting for the pipeline to bring more women into the academy; there is outright discrimination; and so forth [*Holmes et al.*, 2008]. There may be empirical evidence to support some of these reasons but not all. Categorizing them using a sociological framework allows us to test their validity more systematically and then target potential barriers in the appropriate context, rather than randomly trying remedies and hoping that they work.

In the social sciences and among ADVANCE institutions, taxonomies for these proposed reasons, or potential barriers to advancement, abound. We chose a relatively simple one that helps classify these potential barriers for underrepresented faculty. Laursen and coworkers offer an alternative framework in chapter 5.

Risman's [2004] theory of gender categorizes potential barriers into three groups: individual, interactional, and institutional, to explain the underrepresentation of women on the STEM faculty (Figure 3.1).

3.2. Individual Barriers

Individual barriers arise from characteristics of the woman or man. For women geoscientists, we hear, for example: she lacks math ability, she chooses not to pursue an academic career, she prefers nurturing fields or working with people, she chooses to have a family, she lacks the assertiveness to ask for mentorship or to negotiate. We have also heard "she will get cold in polar regions" as a reason to not allow women students to do field work there based on the assumption that women have a lower tolerance of cold due to their biology. Individual barriers are best addressed by better education, mentorship, and advocacy for women. *Professional development programs for women* that clarify the job application process, negotiation, mentorship, networking and its expansion, and the performance review processes (such as for annual evaluations and for tenure and promotion) all address the *individual*. Some individual issues are perceived as not fixable, such as women's personal choices, lower math abilities, or tolerance for cold climates. For some geoscientists (both men and women), individual issues imply a ceiling for the proportion of women we can ever expect in the science: there are simply some women who don't "have it" or don't want it. We address barriers and strategies to overcome individual barriers that are amenable to change in chapters 9 to 15 of this volume.

3.3. Interactional Barriers

Interactional barriers arise from less-than-helpful interactions between and among people. For example, we may assign students to different tasks based on their gender rather than ability: during fieldwork, men students may be assigned to gather the wood while women students are assigned to fix the meal. We hold many assumptions about our roles based on our gender that we've never really examined consciously [e.g., *Valian*, 2004], sometimes for lack of a vocabulary. Such assumptions are *implicit assumptions*. When implicit assumptions have a stifling effect on others, they become *implicit biases*. These differ from overt discrimination (another type of interactional barrier) in that they lurk beneath our conscious awareness. Implicit assumptions arise from our shared culture and belong to members of the culture regardless of gender, race, or ethnicity. Men and women of a given culture share the same implicit assumptions. Rather than encouraging blame or recrimination for holding implicit assumptions, on the contrary, understanding where implicit assumptions come from frees us to examine these assumptions without guilt, since all of us are guilty of holding them. We discuss this topic more fully in chapter 9.

Interactional issues are best addressed by an analysis of underlying assumptions and by mentorship and professional development programs for all faculty. Some of these assumptions may hold up for some individuals, and some break down upon examination. They may arise from stereotypes and do not apply to *all* men or all women but may apply to a few individuals. Let us revisit the example of suitability of women for polar fieldwork discussed above, but now through the interactional lens: the fact that many women successfully do fieldwork in polar regions suggests that this *perception* of women's suitability, the implicit assumption that she is not suitable or capable of polar fieldwork, is an *interactional* barrier, not an individual one inherent in all women. The perception that women cannot handle the cold belongs to the perceiver, not to the woman. That is, a PI may perceive a woman student as more suited to lab work or to base camp work rather than strenuous work on ice. The perception of one person on another's ability is an example of an interactional barrier. If the perception of the PI changed, the woman student could choose to go to the ice and further her career through that experience. Examination through the interactional lens helps us test the validity of our implicit assumptions about each other.

3.4. Institutional Barriers

Institutional barriers include policies, informal practices, and the physical layout of the institution. You may well wonder, "How can an institution be gendered or be uneven in its treatment of faculty based on gender? Isn't an institution, by definition, not a human?" Of course, institutions are created by humans, and the institution's policies and practices reflect those humans' values. Just as corporations within the U.S. have different corporate cultures based on the behavior and expectations of the people at the top, administrators and leaders at academic institutions strongly influence academic culture [*Gappa et al.*, 2007]. Who is at the top in academia? Men make up 80% of the faculty (most at full professor rank) and 85% of administrative positions [*Nelson*, 2007; *Bilen-Green et al.*, 2008]. This male majority has inherited, adopted, and enacted policies without realizing the gendered impact.

Policies that appear to be gender neutral may in fact disadvantage women more than men [*Currie et al.*, 2002; also, see nine studies cited in *Bird*, 2011]. Does the institution offer childcare to its faculty, including sick childcare? Does it allow flexibility in the length of the tenure clock? Are there written policies for part-time tenure-leading appointments or non-tenure-leading appointments? Are lactation rooms available with all the necessary accessories? Do policies or job advertisements, by their wording, generate a chillier climate for women than for men? Do women faculty have access to the same resources as men faculty? This last question arises from a study conducted in 1999 that drew national attention to the uneven distribution of resources (lab space and financial resources) between men and women at MIT [*MIT*, 1999; *Hopkins, et al.*, 2002]. Are certain awards, honors, and chair-ships based on age of the recipient rather than years since PhD or years since tenure? Faculty who have taken a medical leave may be disadvantaged by age-based policies.

The presence or absence of childcare facilities or a worthy childcare referral service illustrates the gendered impact of an institutional policy and program. There are men faculty who are the primary caregivers for their children who are also disadvantaged by the lack of on-site childcare or referral services, but more women in academia act as primary caregivers for their children than their male colleagues; thus, the lack of childcare facilities or an on-campus referral service disadvantages women more than men at this time [e.g., *Mason and Goulden*, 2004a, b].

Strategies that address institutional barriers include the adoption and enactment of new policies that address gender differences. Policies will not always be equal: men will never need lactation facilities, but women do, and without them, their lives are more complicated and difficult than their nonlactating male colleagues. Women must bear children by a biologically determined age. The tenure clock overlaps the biological clock [*de Wet et al.*, 2002], so equitable treatment for women who choose families along with their scientific career, as many men choose, includes flexibility of the tenure clock. These policies and further strategies to make the institution gender neutral are discussed more fully in chapters 7 and 8.

3.5. Multiple Frames

Barriers classified into these three categories can overlap (Figure 3.1): the perception that women choose not to enter academia (an *individual* issue) may hold up to scrutiny for some individuals, but some women choose not to enter the academy because no one has encouraged them to do so or because they perceive a chilly or family-unfriendly environment. This absence of mentorship is an *interactional* issue. Some women choose not to enter the academy because they see the lack of institutional support for families. Lack of childcare support and flexible tenure clocks are *institutional* issues. Thus, "choice" may have institutional, interactional, and individual components. Uneven distribution of resources may arise in part from women's not negotiating as aggressively as men (individual), but the majority's perception of the minority population's worth also plays a role (interactional), as does institutional oversight via ombudspersons and regular salary reviews (institutional). Strategies that address both individual and interactional barriers are discussed further in chapters 9 to 15.

REFERENCES

Bilen-Green, C., K. A. Froelich, and S. W. Jacobson (2008), The prevalence of women in academic leadership positions, and potential impact on prevalence of women in the professorial ranks, *WEPAN Conference Proc.*, http://www.ndsu.edu/fileadmin/forward/documents/WEPAN2.pdf. Accessed Sept. 5, 2012.

Bird, S. (2011), Unsettling universities' incongruous, gendered bureaucratic structures: A case-study approach, *Gender, Work & Organization*, 18(2), 202–230.

Currie, J., B. Thiele, and P. Harris (2002), *Gendered Universities in Globalized Economies*, Lexington Books, New York.

de Wet, C. B., G. M. Ashley, and D. P. Kegel (2002). Biological clocks and tenure timetables: Restructuring the academic timeline, *GSA Today*, *12*, 24.

Gappa, J. M., A. E. Austin, and A. G. Trice (2007), *Rethinking Faculty Work*, Josey-Bass, J. Wiley & Sons, San Francisco, pp. 145–190.

Holmes, M.A., S. OConnell, C. Frey, and L. Ongly (2008), Gender imbalance in US geoscience academia, *Nature Geoscience*, *1*(2), 79–482.

Hopkins, N., L. Bailyn, L. Gibson, and E. Hammonds (2002), *The Status of Women Faculty at MIT: An Overview of Reports from the Schools of Architecture and Planning; Engineering; Humanities, Arts, and Social Sciences; and the Sloan School of Management*, Massachusetts Institute of Technology, Cambridge, MA.

Mason, M.A. and M. Goulden (2004a), Do babies matter? Part II: Closing the baby gap: Do academic careers curb the ability of faculty to form families? *Academe Online*, *90*(6). http://www.aaup.org/AAUP/pubsres/academe/2002/ND/Feat/Maso.htm. Accessed September 5, 2012.

Mason M.A. and M. Goulden (2004b), Marriage and baby blues: Redefining gender equity in the academy, *Annals of the American Academy of Political and Social Science*, *596*, 86–103.

Massachusetts Institute of Technology (1999). A study on the status of women faculty in science at MIT. *The MIT Faculty Newsletter*. http://web.mit.edu/fnl/women/women.pdf. Accessed September 5, 2012.

Nelson, D. (2007), *A National Analysis of Minorities in Science and Engineering Faculties at Research Universities*. http://cheminfo.ou.edu/~djn/diversity/Faculty_Tables_FY07/07Report.pdf. Last accessed 19 Nov 2007.

Risman, B. J. (2004), Gender as a social structure: Theory wrestling with activism. *Gender & Society*, *18*(4), 429–450. doi: 10.1177/0891243204265349

Valian, V. (2004). Beyond gender schemas: Improving the advancement of women in academia. *Natl Women's Studies Assoc. Journal*, *16*(1), 207–220.

SECTION III: SUCCESSFUL STRATEGIES TO ADDRESS THE ISSUE

4

BEST PRACTICES TO ACHIEVE GENDER PARITY: LESSONS LEARNED FROM NSF'S ADVANCE AND SIMILAR PROGRAMS

Mary Anne Holmes

Department of Earth and Atmospheric Sciences, University of Nebraska–Lincoln, Lincoln, Nebraska

ABSTRACT

This chapter introduces the National Science Foundation's (NSF) ADVANCE program, the goal of which is to increase the number of women on the science, technology, engineering, and mathematics (STEM) faculty nationally. Data revealed that women were increasingly earning baccalaureate and advanced degrees in STEM fields, but were not appearing as principal investigators (PIs) on grant applications because so few of them were being hired into academic positions. NSF created serial programs to address this issue, all targeted toward individual women. The 1999 MIT report on gender disparity in STEM departments precipitated discussions that led to the creation of the ADVANCE program by NSF in 2001.

This chapter introduces NSF's ADVANCE program. The goal of this program is to increase the number of women on the STEM faculty nationally. While tracks (types of grants) of this program are used to address individual and interactional barriers to women's advancement, the signature track is "Institutional Transformation" to address institutional barriers to women's participation in STEM fields. ADVANCE addresses barriers to women's entry and advancement in academia through partnerships among STEM faculty, social and behavioral science faculty, experts in organizations and organizational change, campus administrators, and evaluation experts, among others, to generate and test the types of strategies presented in the rest of this volume.

Women in the Geosciences: Practical, Positive Practices Toward Parity, Special Publications 70.
First Edition. Edited by Mary Anne Holmes, Suzanne OConnell, and Kuheli Dutt.
© 2015 American Geophysical Union. Published 2015 by John Wiley & Sons, Inc.

4.1. The National Science Foundation's ADVANCE Program

NSF is charged by Congress to track the up-and-coming STEM workforce. To this end, NSF's National Center for Science and Engineering Statistics has tracked the number of students who annually receive bachelor's, master's and doctorate degrees by STEM discipline since 1966 (www.nsf.gov/statistics/). Data on the students' race, ethnicity, gender, career aspirations, and postgraduate job status are available for selected years.

Examination of these data revealed that women were increasingly earning baccalaureate and advanced degrees in STEM fields but were not appearing as principal investigators on grant applications because so few of them were being hired into academic positions. NSF created serial programs to address this issue, all targeted towards individual women. These included Faculty Awards for Women, Visiting Professorships for Women, Career Advancement Awards, and POWRE Awards (Professional Opportunities for Women in Research and Education) [*Rosser*, 2004].

In 1999, the Massachusetts Institute of Technology (MIT) faculty published a report of a faculty committee that had studied the treatment of men and women and distribution of resources among them at MIT. The committee reported that pretenured women tended to feel that they were treated the same as their male colleagues. Women with tenure who had been faculty longer than their pretenure colleagues saw unequal treatment and outright discrimination, which the committee documented. From the MIT report [*MIT*, 1999]:

> Within three departments the Committee obtained evidence of subtle differences in the treatment of men and women faculty, evidence of exclusion, and, in some cases, evidence of apparent discrimination against women faculty. The Committee documented differences in salary in the recent past, in amount of nine-month salary paid from grants, in access to space, resources, and inclusion in positions of power and administrative responsibility within departments or within the broader MIT community. Differences resulted in women having less or in their being excluded from important professional opportunities. Interviews with women faculty revealed the tremendous toll that exclusion and marginalization take on their professional and personal lives. Problems appear to increase progressively as women approach the same age as their administrators. The Committee believes that problems flourish in departments where non-democratic practices, including administrative procedures whose basis is known only to a few, lead inevitably to cronyism and unequal access to the substantial resources of MIT.
>
> The Committee learned that the percent of women faculty in the School of Science has not increased for at least a decade. As of 1994 there were 22 women faculty, 252 male faculty.

Because of the rigorous data to support these contentions, the MIT administration moved quickly to address them. In January 2001, MIT hosted a meeting of presidents, provosts, chancellors, and 25 women scientists from the nation's top nine universities to discuss the issues raised in the MIT report on the hostile climate for women scientists and engineers at MIT [*Rosser and Lane*, 2002; *Rosser*, 2004].

Based on these results and from analysis of the impact of previous programs, NSF established the ADVANCE program in 2001 to address the underrepresentation of women on the STEM faculty. While prior programs were designed to help individual women, the bulk of the funding for the ADVANCE program has gone to institutions as Institutional Transformation awards to address the institutional barriers that, if removed, could improve hiring, retention, and climate for STEM women faculty. The focus shifted from "fixing the woman" to "fixing" the culture of academia. NSF has provided over $130 million to ADVANCE programs in 41 states, the District of Columbia, and Puerto Rico (NSF Web site: http://www.nsf.gov/crssprgm/advance/index.jsp).

ADVANCE programs to individuals (Leadership awards, PAID awards, ADVANCE Fellowships) and to institutions have engaged men and women scientists, mathematicians and engineers, sociologists, economists, psychologists, and people with expertise in organizations, management, and evaluation. More than a decade of research on what works and what doesn't work, on progress made, and on continuing barriers has yielded a rich literature published in diverse venues (see references). Some results from early institutional awardees are summarized in *Bilimoria* and others [2008] and *Bilimoria and Liang* [2011] and by Sandra Laursen and her colleagues in this volume.

In the following chapters, we present demonstrated, evaluated ADVANCE and ADVANCE-like programs that have increased the number, retention, and promotion of women in the geosciences. There may be programs involving geoscientists that we do not present in this volume, either because we were not aware of them at the time of publication or the person involved did not have time to participate. We apologize for any omissions. The next set of chapters discuss the practices and policies that institutions have adopted to shift their cultures to promote gender equity. The remaining chapters discuss programs that individuals have undertaken to increase retention of women in geoscience.

REFERENCES

Bilimoria, D. and X.F. Liang (2011), *Gender Equity in Science and Engineering: Advancing Change in Higher Education*, Routledge Studies in Management, Organizations and Society, New York.

Bilimoria, D., S. Joy, and X. F. Liang (2008), Breaking barriers and creating inclusiveness: Lessons of organizational transformation to advance women faculty in academia science and engineering, *Human Resource Management*, 47(3), 423–441.

Massachusetts Institute of Technology (MIT) (1999), A study on the status of women faculty in science at MIT, *The MIT Faculty Newsletter*, http://web.mit.edu/fnl/women/women.pdf. Accessed September 5, 2012.

Rosser, S. and E. O. Lane (2002), Key barriers for academic institutions seeking to retain female scientists and engineers: Family-unfriendly policies, low numbers, stereotypes, and harassment, *Journal of Women and Minorities in Science and Engineering, 8,* 161–189.

Rosser, S. (2004), Using POWRE to ADVANCE: Institutional barriers identified by women scientists and engineers, *Natl Women's Studies Assoc J, 16*(1), 50–78. doi: 10.1353/nwsa.2004.0040.

SECTION III.A: INSTITUTIONAL STRATEGIES

5

STRATEGIC INSTITUTIONAL CHANGE TO SUPPORT ADVANCEMENT OF WOMEN SCIENTISTS IN THE ACADEMY: INITIAL LESSONS FROM A STUDY OF ADVANCE IT PROJECTS

Sandra L. Laursen,[1] Ann E. Austin,[2] Melissa Soto[2] and Dalinda Martinez[2]

[1]*Ethnography & Evaluation Research, University of Colorado Boulder, Boulder, Colorado*
[2]*Higher, Adult, and Lifelong Education, Michigan State University, East Lansing, Michigan*

ABSTRACT

This chapter looks at initial lessons from a study of ADVANCE-IT projects. We take a cross-institutional, analytical, and synthetic approach to extract useful lessons about organizational strategies that support the success of women scholars in science, technology, engineering, and mathematics (STEM) fields. The study uses a mixed-methods approach and draws upon conceptual frameworks addressing organizational change. Our sample is 19 institutions that received ADVANCE IT awards in Rounds 1 and 2 of grants (2001–2004), and carried out their projects over five to eight years. Using *Bolman and Deal's* [1991] framework, we examine strategic interventions through the structural, human resource, political, and symbolic perspectives. We highlight an approach for identifying and categorizing institutional change interventions and provide examples of change strategies common within our sample. We hope this approach will guide those seeking to enhance women's success in the academy. By viewing their choices as strategic and context-dependent, leaders can more wisely select and combine interventions to craft a systemic change portfolio. We encourage readers to be alert to subsequent publications where we lay out our ideas in depth.

Women in the Geosciences: Practical, Positive Practices Toward Parity, Special Publications 70. First Edition. Edited by Mary Anne Holmes, Suzanne OConnell, and Kuheli Dutt. © 2015 American Geophysical Union. Published 2015 by John Wiley & Sons, Inc.

To solve challenges of the future, diverse and creative perspectives on science and engineering are required. Yet higher education continues to grapple with the underrepresentation of women in the STEM fields (science, technology, engineering and mathematics). Earlier chapters in this volume detail the challenges for women academics at all stages: chilly climate, harassment, exclusion from networks and leadership roles, norms and expectations that privilege a single-minded focus on work and a linear career path, and the coincidence of tenure timelines with child-bearing years. These challenges are faced by faculty across disciplines but are often particularly acute for women in the sciences and engineering, where conflicting gender schemas for woman and scientist, engineer, or professor lead to bias even among well-intentioned colleagues [*Valian*, 1998]. As a result, female faculty in STEM fields may leave academe [*Rosser*, 2004a]. Others remain but report lower job satisfaction as they manage these challenges on a daily basis [*Bilimoria, et al.*, 2006; *Callister*, 2006].

These problems affect the talent pool available to address societal problems today and shape the pipeline of scientists through faculty mentoring of future scholars. Indeed, young women considering academic careers make decisions based on their observations of female faculty and their experiences [*Rice et al.*, 2000; *De Welde and Laursen*, 2011).

Earlier chapters also have detailed the intent and design of the U.S. National Science Foundation's (NSF) ADVANCE program. The program represented a formal shift from viewing women's underrepresentation as a problem of women, "fitting women in" to existing structures and enhancing their professional competitiveness [*Bystydzienski and Bird*, 2006, p. 3], to recognizing that organizational structures and advancement criteria in the academy are optimized for traditional male career patterns. Earlier NSF programs supported the advancement of individual women but did not address the systemic nature and fundamental causes of the challenges academic women face [*Rosser and Lane*, 2002; *Rosser*, 2004b; *Wylie, Jakobsen and Fosado*, 2007]. From its 2001 initiation through 2011, more than 50 ADVANCE Institutional Transformation (IT) grants have been awarded to institutions across the U.S. More than 80 ADVANCE PAID (Partnerships for Adaptation, Implementation, and Dissemination) projects support adoption and adaptation of their strategies in other organizations.

5.1. Study Design and Methods

Our research team is examining the work done at ADVANCE IT institutions, drawing upon the knowledge gained as institutions have experimented, implemented, and adjusted their strategies, and documented their approaches, successes, and challenges. We take a cross-institutional, analytical, and synthetic approach to extract useful lessons about organizational strategies that support the success of women scholars in STEM fields. The central research question is

*What has been learned about the effectiveness and long-term viability of orga-
nizational change efforts to create institutional environments that are conducive to
the success of women scholars, particularly in STEM fields?*

To answer this question, we are exploring several aspects of ADVANCE IT
projects' change strategies, including the following:
- The change strategies chosen by institutions, their initial goals and ultimate
  impact
- The reasons why these strategies have been effective or not
- The impact of organizational culture and context on the initial choice and
  ultimate impact of the chosen strategies
- "Lessons learned" about successful theories of change and effective strategies
  and processes for change.

The study uses a mixed-methods approach and draws upon conceptual frameworks addressing organizational change. Our sample is 19 institutions that received
ADVANCE IT awards in Rounds 1 and 2 of grants (2001–2004) and carried out
their projects over five to eight years. Using a rubric developed from our conceptual
frameworks and guiding literature, we analyzed each project's annual and final
reports, evaluation reports, Web site, and other materials. We interviewed principal
investigators from each campus and prepared a narrative summary of the goals,
objectives, strategies, and outcomes of each institution's work.

Based on this material, five sites were chosen for detailed examination as case
studies. We interviewed more than 170 project participants and leaders at these
sites during 2011–12. Currently, we are analyzing both the broader data set and
the in-depth case studies to identify and categorize individual change interventions, examine how they combine to build a change portfolio, and consider how
change interventions may be selected or adapted to fit a specific institutional
context.

This chapter constitutes an early report of our working hypotheses. We
highlight an approach for identifying and categorizing institutional change
interventions and provide examples of change strategies common within our
sample. We propose these ideas to change leaders as a way to think about these
issues and devise a strategic approach to change for their own institution. We
encourage readers to be alert to subsequent publications where we lay out our
ideas in depth.

5.2. Theoretical and Empirical Perspectives

We view universities as complex organizations composed of multiple, loosely
coupled, interconnected subsystems [*Birnbaum*, 1988; *Bolman and Deal*, 1991;
Cohen and March, 1991; *Weick*, 1976]. Thus, an overall change strategy cannot
depend on a single type of intervention. Yet any particular intervention might be
deployed to serve multiple goals and in a variety of forms that may depend on the

context or system in which it is introduced. Successful change efforts will be nonlinear and will benefit from deliberate efforts to connect an array of strategies [*Eckel et al.*, 2001; *Kezar*, 2001; *Senge*, 1990].

As our primary conceptual framework, we draw upon *Bolman and Deal's* [1991] multiframe model. In this model, four main perspectives serve as viewpoints for examining organizational issues: structural, human resources, political, and symbolic perspectives. All of these perspectives function as frames or "lenses that bring the world into focus" (p. 11) through which to understand organizational issues, in this case, change strategies to advance gender equity.

The *structural* frame emphasizes policy and procedure as tools to improve equity in advancement or address critical junctures in faculty careers. This lens recognizes the importance of formal rules, policies, management hierarchies, and relationships within organizations.

The *human resource* frame emphasizes the importance of the demographics, experiences, needs, and feelings of the people involved in an organization, especially the faculty who are at the center of the institution.

The *political* frame takes into account issues of resource allocation and the sources and seats of power in the university, including both those in formal institutional roles and thought leaders who hold high status on campus.

Finally, the *symbolic* frame focuses attention on issues of meaning and culture within an organization, including the rituals, stories, and celebrated individuals, and the process through which sense-making takes place within the organization [*Eckel et al.*, 2001].

In addition to Bolman and Deal's work, we draw upon prior investigations of ADVANCE activities across multiple institutions. *Stewart* and colleagues [2007] provide a "partial compendium of institutional transformation efforts" (p. 9). Drawing on examples from institutions with ADVANCE IT grants, their book sketches key elements and principles guiding the intervention strategies that have been developed, describes institutional programs and strategies, and highlights institutional cases of implementation.

Studies by Bilimoria and coworkers [*Bilimoria et al.*, 2008; *Bilimoria and Liang*, 2012] and by *Fox* [2008] have gone somewhat further. Each has categorized the activities of ADVANCE IT awardees to draw conclusions about the popularity of particular change activities. *Fox* [2008] analyzed these efforts in light of the literature on transformational change in higher education, suggesting key factors that contribute to transformation, while *Bilimoria* and others [2008] proposed a model linking these efforts to desired change outcomes. *Bilimoria and Liang* [2012] offer a detailed analysis of quantitative data collected by ADVANCE IT institutions to measure progress in hiring, retention, and promotion of faculty. Our work promises to yield more detailed insights into the hows and whys of institutional transformation and the factors that may explain why and how change does or does not occur within specific contexts.

5.3. Findings

We highlight each of Bolman and Deal's lenses in turn and provide examples from our data of strategic interventions used by ADVANCE IT institutions. Our ongoing work addresses each of the interventions in greater detail and examines how they operate within a change portfolio.

5.3.1. Structural Strategic Interventions

Structural strategic interventions focus on identifying and improving formal policies and procedures. Because institutional policies and traditions have been historically developed in response to male career patterns, policies and practices may unintentionally inhibit women's advancement or success in the academy. Examples of structural strategic interventions include the following.

5.3.1.1. Analysis, tracking, and revision of tenure and promotion policies. Policies may be reviewed and revised to make them more transparent and flexible. For example, a policy for tenure-clock extension that allows faculty extra time before the tenure review after a major life event such as childbirth may be extended automatically to all faculty who are new parents. The change seeks to encourage policy use and make tenure-clock extension the norm rather than the exception.

Other structural interventions use administrative or faculty roles to carry out particular duties. For example, a tenure ombudsperson may be appointed to serve as a neutral advocate for each tenure candidate. The ombudsperson ensures that institutional review processes are carried out in accordance with written procedures and encourages fair discussion focused on the candidate's qualifications. This structure thus addresses real and perceived fairness in tenure review.

5.3.1.2. Creation, dissemination, and monitoring of the use of work/life policies. Policies may set conditions and terms under which faculty can arrange modified duties in response to personal or family needs, or take family leave. Research has shown that the mere existence of a family leave policy, for instance, is not sufficient for it to be used by those whom it is intended to benefit [*Drago et al.*, 2005]. The U.S. context of minimal work-family policy at the federal level [*Hegewisch and Gornick*, 2008] leads to disparate initial situations at various ADVANCE IT institutions; thus, some institutions make progress merely by smoothing pathways for faculty to take unpaid leave under the provisions of the U.S. Family and Medical Leave Act, while others work to adapt existing institutional policies for paid leave to accommodate a broader range of life events.

5.3.1.3. Tools and training to enhance recruitment and hiring of women faculty. Several ADVANCE institutions have developed tool kits or training for search committees, targeting both active recruitment of a diverse applicant pool and fair evaluation of applications. These often include research-based information on implicit bias and how it may affect the review of job application materials. Implicit bias emerges from unintentional application of common mental schemas. For example, societal schemas for "women" conflict with those for "engineer," and thus it is more difficult to imagine how women can also be good engineers [*Valian*, 1998]. Recommendation letters are written differently for men and women [*Trix and Psenka*, 2003; *Schmader et al.*, 2007]. Awareness of this body of research can help search committee members to become alert to their own biases and to counter them in discussions (see also chapter 9).

In general, our data suggest that structural intervention strategies can be slow and painstaking to implement. But these interventions have impact well beyond the lifetime of the grant, unlike interventions that end when grant resources are exhausted. While structural interventions may not provide early successes in improving gender equity in an institution, they can have a sizable impact as part of the overall change portfolio.

5.3.2. Human Resource Interventions

Human resource interventions attend to the characteristics, experiences, needs, and aspirations of the institution's people, especially the faculty, who are at the core of the institution. Such interventions typically provide professional development, resources, and opportunities for people to expand their own knowledge, skills, and capacities.

5.3.2.1. Professional development of work skills for career success. Professional development activities developed at ADVANCE institutions focus on enhancing faculty members' ability to do their jobs by providing information and building skills. Short "one-off" sessions or more intensive offerings may address grant writing, managing people, money and time, negotiation, networking, self-promotion, and long-term career planning. Target audiences may include early-career faculty pursuing tenure, or mid-career faculty balancing multiple responsibilities. Overall, faculty development activities were a primary means by which projects addressed retention of STEM women faculty.

5.3.2.2. Mentoring, coaching, and networking. ADVANCE institutions have offered a wide range of mentoring and coaching programs, ranging from formal individual mentoring of early-career faculty through informal peer mentoring and on-demand coaching to troubleshoot specific problems. Some programs focus on pretenure faculty; others support department chairs or midcareer faculty pursuing advancement. Networking activities tended to be informal and organized around a meal or a presentation. ADVANCE projects often leveraged workshops or celebrations to include networking.

5.3.2.3. Small grants. ADVANCE IT institutions hosted a variety of grants programs. Some programs focused on the scholarly development of faculty, especially early in their careers. Formats that encouraged faculty to develop on- or off-campus collaborations appeared to have especially high return on investment, yielding presentations, publications, and successful applications for external funding. Grants to support career and life transitions typically offered support for faculty to resume or revitalize scholarly activity after a life event or career transition. Funds might support research travel, assistance, or equipment, family care during fieldwork or conferences, or professional development to learn a new skill or make new contacts. These programs too had high bang for the buck, both in terms of faculty productivity and in recipients' feelings of being supported and valued by the institution.

In general, human resource interventions have been common among ADVANCE IT projects. Relatively easy to implement, they meet direct needs of individual women faculty and others. Many institutions have found creative ways to institutionalize elements of these programs.

5.3.3. Political Strategic Interventions

Political strategic interventions emphasize issues of leadership, power, and resource allocation.

5.3.3.1. Leadership development for deans, chairs, and committee leaders. These interventions recognize the crucial formal and informal roles of deans and chairs in allocating resources, reviewing personnel, establishing priorities and procedures, and setting a tone. Enhancing these leaders' skills, capacities, and awareness is seen as a high-leverage strategy for influencing the success and satisfaction of all faculty.

5.3.3.2. Use of the faculty governance system to enact policy change. Several projects made good use of their local governance systems to call for policy changes or reviews; to discuss, publicize, and enact proposed policies; and to garner broad support among the faculty. ADVANCE leaders drew upon their previous experiences as senior faculty and institutional leaders to devise these governance-related approaches.

5.3.3.3. Policy or action committees. ADVANCE projects made use of committees and task forces to accomplish specific tasks, share information, engage interest, and build visibility. Project leaders might share data, ask for committee advice or input, and provide white papers. They sought to place themselves on influential committees or were invited to participate by a president, provost, or dean.

5.3.3.4. Institutional data-gathering and dissemination. ADVANCE institutions used the indicator data gathered for NSF to identify particular institutional issues, highlight progress, and raise awareness. Some projects provided data to each school or department on the demographics of its faculty in comparison with national samples. At some campuses, such practices generated a constituency for the data, leading to the institutionalization of data collection and publishing systems at the grant's end. Projects also used research data to point out concerns and raise knowledge, as in implicit bias training for search committees. Internal evaluation and research studies were used to identify problems and needs, generate interest and visibility, and document progress for the institution and the funder.

Overall, we observed that political strategic interventions could be influential, though difficult to implement in practice. In the best cases, ADVANCE leaders had cultivated politically powerful allies, including institutional leaders in formal roles but also high-status faculty who served as opinion leaders or "organizational catalysts" [*Sturm*, 2007]. Turnover in the upper administration could change the political landscape in unanticipated ways, and developing effective political strategies required time and persistence.

5.3.4. Symbolic Strategic Interventions

Symbolic strategic interventions emphasize issues of meaning within the organization.

5.3.4.1. Publicity and communications. ADVANCE leaders maintained a steady drumbeat of communication to inform campus stakeholders, solicit participation, and engage interest, through their Web site, newsletters, and participation in meetings with academic and administrative units and committees. It took time, energy, and repetition to insert ADVANCE into the general consciousness, and it was not always immediately obvious whether these efforts were having any benefit. In addition to communications from the ADVANCE office, a message from upper institutional leaders could have high symbolic value: for example, when a university president highlighted ADVANCE in his or her annual "state of the campus" report. Such instances could help to articulate and shift institutional values and cultures over time.

5.3.4.2. Awards and recognition. Campus and external awards were used to celebrate women's accomplishments and raise awareness of women's contributions. Some projects targeted and nominated specific women; others shared information on awards and how to prepare a nomination. Some developed internal awards to recognize women STEM scholars on their campus. Such recognition could help create bridges to influential senior women who could become allies of ADVANCE.

5.3.4.3. Special events. Many projects used visiting scholars to offer special events on campus. A lecture by an eminent scholar in his or her field might garner publicity and draw in faculty who would not otherwise engage in ADVANCE.

Meals and receptions for the visitor could serve to network faculty, introduce young scholars to a distinguished senior colleague, and engage department heads and deans in ADVANCE. Other special events included annual celebrations of women's accomplishments, receptions to honor ADVANCE grantees, and ribbon-cutting events.

In general, symbolic strategic interventions were viewed as having good value over the long term, even if their short-term effectiveness was less evident. Special events especially offered high visibility and credibility that could lay the groundwork for some types of political interventions.

5.4. Discussion

So far we have analyzed interventions as primarily viewed through a single lens. Yet many interventions may be viewed through multiple frames. For example, a special ADVANCE-sponsored lecture might have high symbolic value but also pull faculty together to network individuals, serving thus as a human resource intervention. Inviting the president to introduce the visiting scholar might add a political opportunity. In future analyses, we will analyze how interventions can be viewed through different lenses to reveal affordances or limitations less evident under a single lens.

Moreover, while our research study focuses on institutions, our framework for analyzing strategic interventions applies as well to other types of ADVANCE projects. For example, the Earth Science Women's Network (ESWN, www.eswnonline.org; see also chapter 14) is a network of early-career geoscientists seeking to connect women and share knowledge of options for how to navigate a career. While ESWN takes primarily a human resource approach in seeking to enhance women's knowledge, skills, and confidence, this approach also has political ramifications, as women are empowered to improve their situations. As an example, an online discussion of child care at professional meetings fostered a connection to someone who could help to resolve child care concerns at one society's meeting.

In another example, the AWARDS project led by the Association for Women in Science (http://www.awis.org/displaycommon.cfm?an=1andsubarticlenbr=397) has carried out analyses of awards given by several professional societies. As a study collaborator, the American Geophysical Union is now reviewing recommendations to increase women's representation among the winners of society awards. For an awardee, this strategy may be viewed as a human resource intervention, while within the society, it has high symbolic and political value.

5.5. Summary

We have sketched the strategic interventions used by ADVANCE IT projects to create institutional environments that support the success of women STEM scholars. Using *Bolman and Deal's* [1991] framework, we are examining strategic interventions through the structural, human resource, political, and symbolic perspectives. This analysis across institutions will be complemented by analyses of how multiple interventions are combined, through the in-depth case studies. We hope this approach will guide those seeking to enhance women's success in the academy. By viewing their choices as strategic and context-dependent, leaders can more wisely select and combine interventions to craft a systemic change portfolio.

ACKNOWLEDGMENTS

We are grateful to the leaders at the 19 study sites who responded to our queries and participated in interviews, and those at five sites who graciously hosted us for in-depth case studies. We thank the members of our external advisory board and an ad hoc working group for their advice and insights. This work was supported by an ADVANCE PAID grant from the National Science Foundation (HRD-0930097). All opinions and ideas are the responsibility of the authors.

REFERENCES

Bilimoria, D., S. Joy, and X. Liang (2008), Breaking barriers and creating inclusiveness: Lessons of organizational transformation to advance women faculty in academic science and engineering, *Human Resource Management*, *47*(3), 423–441.

Bilimoria, D., and X. Liang (2012), *Gender Equity in Science and Engineering: Advancing Change in Higher Education*, Taylor and Francis, New York.

Bilimoria, D., S. R. Perry, X. Liang, E. P. Stoller, P. Higgins, and C. Taylor, (2006), How do female and male faculty members construct job satisfaction? The roles of perceived institutional leadership and mentoring and their mediating processes, *Journal of Technology Transfer*, *31*(3), 355–365.

Birnbaum, R. (1988), *How Colleges Work*, Jossey-Bass, San Francisco.

Bolman, L. G., and T. E. Deal (1991), *Reframing Organizations: Artistry, Choice, and Leadership*. Jossey-Bass, San Francisco.

Bystydzienski, J. M., and S. R. Bird (eds.) (2006), *Removing Barriers: Women in Academic Science, Technology, Engineering, and Mathematics*, Indiana University Press, Bloomington.

Callister, R. R. (2006), The impact of gender and department climate on job satisfaction and intentions to quit for faculty in science and engineering fields, *Journal of Technology Transfer*, *31*(3), 367–375.

Cohen, M. D., and J. G. March (1991). The process of choice. In M. W. Peterson, E. E. Chaffee, and T. H. White (eds.), *Organization and governance in higher education: An ASHE reader* (4th ed., pp. 175–181), Simon and Schuster, Needham Heights, MA.

De Welde, K., and S. L. Laursen (2011), The glass obstacle course: Informal and formal barriers for women Ph.D. students inSTEM fields, *International Journal of Gender, Science and Technology*, *3*(3), 571–595.

Drago, R., C. Colbeck, K. D. Stauffer, A. Pirretti, K. Burkum, J. Fazioli, G. Lazarro, and T. Habsevich (2005), Bias against caregiving, *Academe*, *91*(Sept.-Oct.), 22–25.

Eckel, P., M. Green, and B. Hill (2001), Riding the waves of change: Insights from transforming institutions, *On Change V: An occasional paper series of the ACE Project on Leadership and Institutional Transformation and the Kellogg Forum on Higher Education Transformation,* American Council on Education, Washington, D.C.

Fox, M. F. (2008), Institutional transformation and the advancement of women faculty: The case of academic science and engineering. In J. C. Smart, ed., *Higher education: Handbook of Theory and Research*, vol. *23*, pp. 73–103, Springer, New York.

Hegewisch, A., and J. C. Gornick (2008), *Statutory Routes to Workplace Flexibility in Cross-National Perspective*, Institute for Women's Policy Research, Washington, D.C.

Kezar, A. (2001). Understanding and facilitating organizational change in the 21st century: Recent research and conceptualizations, *ASHE-ERIC Higher Education Report*, *28*(4), Jossey-Bass, San Francisco.

Rice, R. E., M. D. Sorcinelli, and A. E. Austin (2000), *Heeding new voices: Academic careers for a new generation*, American Association of Higher Education, Washington, D.C.

Rosser, S. V. (2004a), *The Glass Ceiling: Academic Women Scientists and the Struggle to Succeed*, Routledge Press, New York.

Rosser, S. V. (2004b), Using POWRE to ADVANCE: Institutional barriers identified by women scientists and engineers, *NWSA Journal*, *16*(1), 50–78.

Rosser, S. V., and E. O. Lane (2002), Key barriers for academic institutions seeking to retain female scientists and engineers: Family-unfriendly policies, low numbers, stereotypes, and harassment, *Journal of Women and Minorities in Science and Engineering*, *8*, 163–191.

Schmader, T., J. Whitehead, and V. H. Wysocki (2007), A linguistic comparison of letters of recommendation for male and female chemistry and biochemistry job applicants, *Sex Roles*, *57*(7–8), 509–514.

Senge, P. M. (1990), *The Fifth Discipline: The Art and Practice of the Learning Organization*, Doubleday, New York.

Stewart, A. J., J. E. Malley, and D. LaVaque-Manty (eds.) (2007), *Transforming Science and Engineering: Advancing Academic Women*, University of Michigan Press, Ann Arbor.

Sturm, S. (2007), Gender equity as institutional transformation: The pivotal role of "organizational catalysts." In A. J. Stewart, J. E. Malley, and D. LaVaque-Manty (eds.), *Transforming Science and Engineering: Advancing Academic Women* (pp. 262–280), University of Michigan Press, Ann Arbor.

Trix, F., and C. Psenka (2003) Exploring the color of glass: Letters of recommendation for female and male medical faculty, *Discourse and Society*, *14*(2), 191–220.

Valian, V. (1998), *Why So Slow? The Advancement of Women*, MIT Press, Cambridge, MA.

Weick, K. W. (1976), Educational organizations as loosely coupled systems, *Administrative Science Quarterly*, *21*, 1–19.

Wylie, A., J. R. Jakobsen, and G. Fosado (2007), *Women, Work and the Academy: Strategies for Responding to "Post-Civil Rights Era" Gender Discrimination*, New Feminist Solutions, Barnard Center for Research on Women, New York.

6

INSTITUTIONAL TRANSFORMATION: THE LAMONT-DOHERTY EARTH OBSERVATORY EXPERIENCE

Kuheli Dutt

Lamont-Doherty Earth Observatory, Columbia University, Palisades, New York

ABSTRACT

In recent years there has been a huge body of literature examining the underrepresentation of women in STEM fields. Causes cited range from stereotype threat and subconscious gender bias to social and environmental factors. Historically men have tended to dominate STEM fields, but in recent years women's participation has increased significantly. For example, in the earth, atmospheric, and ocean sciences, only 3% of the doctoral degrees in 1966 were awarded to women; that number increased to approximately 35% in 2006. Yet despite these increases women remain underrepresented in STEM faculty across the country. This chapter provides an overview of the institutional transformation experience of Columbia University's Lamont-Doherty Earth Observatory over the period 2005–2012. The focus of these efforts has been the integration of diversity considerations into all academic decision-making processes, and bringing about a systemic change in institutional procedures. The results show an increase in women across all scientific ranks, with the most significant increase at the Doherty Scientist/Lamont Assistant Research Professor rank, up from 18% women in 2005 to 39% women in 2012. Many of the issues discussed in this chapter are not unique to LDEO but are common to other scientific institutions as well, suggesting that the LDEO experience could serve as a model for peer institutions facing similar problems.

Women in the Geosciences: Practical, Positive Practices Toward Parity, Special Publications 70.
First Edition. Edited by Mary Anne Holmes, Suzanne OConnell, and Kuheli Dutt.
© 2015 American Geophysical Union. Published 2015 by John Wiley & Sons, Inc.

6.1. Introduction and Overview

In recent years there has been a huge body of literature examining the under-representation of women in science, technology, mathematics and engineering (STEM) fields. Causes cited range from stereotype threat and subconscious gender bias to social and environmental factors. Historically men have tended to dominate STEM fields, but today women's participation has increased significantly. For example, in 1966 women earned only 12% of the doctoral degrees in the biological sciences and about 2% of the doctoral degrees in physics; those numbers increased to 48% in the biological sciences and 17% in physics in 2006 [NSF, 2008]. In the earth, atmospheric, and ocean sciences, only 3% of the doctoral degrees in 1966 were awarded to women; that number increased to approximately 35% in 2006 [NSF, 2009]. In 2011, at Columbia University, women made up approximately 59% of the doctoral students in earth and environmental sciences, with men at approximately 41%.

Yet despite these increases, women remain underrepresented in STEM faculty across the country. According to 2006 data on STEM faculty in four-year educational institutions, only 7.2% of tenured faculty in engineering and 13.7% of tenured faculty in the physical sciences were women, as indicated in a report by the American Association of University Women (AAUW) [Hill et al., 2010]. The report explores various causes of underrepresentation of women among STEM faculty despite women's progress in education in the STEM fields, these causes including cognitive sex differences and beliefs about intelligence, differences in individual preferences between boys and girls, stereotype threat and subconscious gender bias, workplace environment, and family responsibilities. A report by the Center for American Progress, Berkeley Law School, and Berkeley Center on Health, Economic, and Family Security [Goulden et al., 2009] revealed that family formation, most importantly marriage and childbirth, accounts for the biggest leak in the pipeline for women, these leaks being most significant during the postdoctoral years, and the subsequent transition from junior to senior faculty, that is, receipt of tenure. Their findings also showed that married women with children were less likely to enter a tenure-track position after receiving a PhD than married men with children; and if they did enter a tenure-track job, they were also significantly less likely than their male counterparts to obtain tenure. By contrast, single women without children had approximately the same success rates as married men with children in obtaining tenure. This suggests that women bear a disproportionate burden of familial responsibilities, which hinders their career advancement. The report also revealed that scientists often make decisions about their career path while still in training, that is, during the doctoral and postdoctoral years. The majority of doctoral students and postdoctoral scholars surveyed (both male and female) also indicated that they considered research-intensive universities the least family friendly of possible career paths.

To add to the problem, there exists a highly complex legal landscape where affirmative action policies have come under attack in recent years, with institutions of higher education often being the focal point of controversial debates on this issue. Gender- and race-specific programs and policies have come under fire for being discriminatory against the majority. Critics of such programs and policies perceive them as ineffectual, leading to a tradeoff between diversity and academic excellence, while advocates of such policies perceive them as crucial for diversity and for contributing to academic excellence. From *Grutter vs. Bollinger* [2003] to the more recent *Fisher vs. University of Texas* [2011], these cases demonstrate how polarizing and controversial many affirmative action policies can be.

It is against this backdrop, under immense pressure from all sides, that university administrators and leaders must work towards transforming their institutions and addressing the causes and effects of underrepresentation by marginalized groups. Barriers to participation of minority groups need to be addressed without being perceived as discriminatory against majority groups. Causes of such underrepresentation are often historical and reflect structural problems within the institution and more generally, within society. In addition to the legal complexities surrounding affirmative action policies, university leaders need to be aware that simply offering legal protection to underrepresented groups will in and of itself not transform an institution into a more inclusive one. Factors that impact diversity are often the same factors that impact academic affairs, factors such as institutional accountability, transparency, and integrity in decision-making processes. *A systemic change is necessary, with diversity considerations being integrated into core institutional values and all practices and procedures pertaining to academic affairs.* This has been the approach adopted by Columbia University's Lamont-Doherty Earth Observatory (LDEO).

6.2. Institutional Transformation Efforts at LDEO

Loosely following the framework and concepts outlined in *Sturm* [2006], this chapter outlines the institutional transformation experience of Columbia University's LDEO. Before discussing the institutional transformation experience, a brief description of LDEO and its basic structure is in order.

6.3. LDEO Institutional Structure

Lamont-Doherty Earth Observatory is located on Columbia University's Lamont Campus. The Lamont Campus is home to approximately 500 full-time employees (including 200+ scientists) in scientific, staff, administrative, and faculty positions. LDEO operates in a predominantly soft money environment, with scientists raising funds from external (usually federal) agencies to support their research. Scientists at LDEO are assigned to one of five research divisions, each headed by an associate director. The largest group of scientists on the Lamont Campus, the Lamont Research Professors (formerly Doherty Research Scientists), have been the focus of much of LDEO's institutional transformation efforts.[1] The LDEO scientific staff is predominantly white male (typical of the earth sciences), with numbers of women decreasing at senior ranks. The central oversight of all research divisions rests with the director. The primary policy-making body within LDEO is the executive committee, headed by the director, with representation from all LDEO units and divisions. It is this body that provides guidance to the director on policy and programmatic issues pertaining to the present and future directions of LDEO.

6.4. Columbia's ADVANCE IT Grant: Setting the Stage for Institutional Transformation

Columbia University received an ADVANCE IT grant from 2004 to 2009 in the second round of the institutional transformation awards. While diversity issues were already beginning to be discussed at LDEO prior to ADVANCE, the ADVANCE grant acted as a catalyst, bringing to the forefront diversity issues at an institutional level, requiring LDEO to formally examine its track record with respect to diversity. Quantitative and qualitative baselines were established through the administration of a work environment survey carried out by the ADVANCE team in 2005. This survey covered topics such as work environment, professional employment, diversity, work-life, and demographics. The findings of the 2005 report showed the following:

- Approximately 20% of research scientists were women. That proportion further declined as rank increased.
- Women consistently reported experiencing a more difficult work environment than did men.
- They reported receiving less respect from colleagues and perceived departmental/unit processes as less fair than their male colleagues.
- Women reported experiencing gender-related discrimination and adversity.
- Neither the men nor the women surveyed believed that diversity was a goal of their department/unit.
- More than half of the female respondents reported that their family responsibilities hindered their career advancement.

[1]The Department of Earth & Environmental Sciences (DEES) is also located on the Lamont Campus. DEES faculty work closely with LDEO scientists and are part of the same intellectual framework. However the administration of DEES falls under the School of Arts & Sciences, not the LDEO Directorate. Accordingly, the institutional transformation efforts outlined in this chapter pertain to LDEO experience.

Regardless of gender, a significant proportion of scientists had been approached with outside offers, and both women and men were equally likely to translate an outside offer into a retention offer. There were also no significant differences in the work activities in which scientists were engaged. Men and women also reported similar participation as members and chairs of institutional committees as well as similar levels of involvement in national or international committees. Also, responses to the open-ended questions indicated that both men and women found the LDEO environment to be intellectually stimulating.

The recommendations of the ADVANCE committee for LDEO were to
- Improve the institutional climate by promoting awareness of subconscious bias and stereotype threat,
- Promote a more diverse work environment,
- Adopt family-friendly policies,
- Create incentives and accountability for mentoring, and
- Increase the number of women among the scientific staff and in leadership positions.

6.5. Creation of Office of Academic Affairs and Diversity

Building on the research and recommendations of the CU ADVANCE team and the feedback of the LDEO scientific staff, and recognizing that diversity issues are inextricably intertwined with academic affairs (such as appointments and promotions, salary structures, and institutional governance), the LDEO Directorate created the Office of Academic Affairs and Diversity (OAAD) in 2008, which aims at creating a culture of inclusiveness and removing barriers to full participation, via an ongoing process of institutional transformation. This office has the following goals:
- To increase the recruitment and retention of women and minorities among the scientific staff,
- To foster the career advancement of junior scientists and postdoctoral scholars, and
- To create a culture of inclusiveness and to improve the quality of the work environment for all by promoting transparency and institutional accountability.

In order to facilitate the achievement of these goals, this office has access to the highest levels of leadership within LDEO and is housed within the LDEO Directorate. It is responsible for developing policies, programs, and initiatives impacting academic affairs and diversity issues, and participates at all levels of decision making and governance within LDEO (including the LDEO Executive Committee, the primary policy-making body), with the goal of including diversity discussions into all academic decision-making processes. This office acts as an "organizational catalyst" [see *Sturm*, 2006], operating at the convergence of different domains. This office is strategically placed at the convergence of academic decision-making processes at LDEO and multiple levels of institutional activity pertaining to diversity, and more generally, the quality of the work environment.

6.6. Understanding Factors Impacting Diversity

Before one can adequately address the barriers to full participation, one must first understand the causes of those barriers. In STEM disciplines, "diversity" has commonly referred to gender and racial diversity, with these two indicators often being grouped together in diversity discussions. In reality the circumstances surrounding these two indicators of diversity vary greatly; therefore, solutions that work for one may not necessarily work for the other, a result consistent with the LDEO experience.

With respect to gender diversity, one of our main challenges is to find ways to arrest the leak in the pipeline and to retain and advance women in science. At LDEO in 2005, approximately 50% of the postdoctoral scientists were women; yet at the very next level (the Doherty Associate Research Scientist level) only about 18% were women, indicating a significant leak in the pipeline. This suggests that an available pool exists, and that by adopting progressive and proactive policies, it is possible for an institution to increase the number of women in its ranks. Underrepresentation of women in the earth sciences is not so much a pipeline problem as an institutional one, with its roots in historical, social, and cultural factors. As indicated earlier in the AAUW report, these factors include cognitive sex differences and beliefs about intelligence, differences in individual preferences between boys and girls, stereotype threat and subconscious gender bias, workplace environment, and family responsibilities.

With respect to racial diversity, however, the scenario is very different. National-level data indicate that approximately 88% of all doctoral degrees awarded in the earth sciences in 2010 were to Caucasians; and that Blacks, Hispanics, and Native Americans[2] *taken together* accounted for approximately 5% of all doctoral degrees in the earth sciences [*U.S. Census Bureau*, 2011]. In other words, there is a severely limited pipeline, this reflecting the situation in middle and high schools across the country, something that is typically beyond the immediate reach of individual universities. The LDEO experience has been consistent with national-level data, with the scientific staff being predominantly (approximately 89%) white. Racial minorities (i.e., nonwhite) at LDEO are mostly foreign-born Asians and Hispanics, with very little representation from U.S. underrepresented groups.

---

[2]NSF categorizes Blacks, Hispanics, and Native Americans as "underrepresented minorities" (or URM) in STEM disciplines.

The case for diversity in the sciences is rooted in the need for sustained scientific advancement. There are two aspects to this. One, to retain its leadership position in science and technology, the United States needs skilled scientists and engineers to enter the workforce. With women constituting almost 50% of the labor force and underrepresented minorities projected to constitute approximately 30% of the U.S. population by 2020 [*Nelson*, 2007], a proactive approach is needed to include these groups in research universities if the United States wants to maintain its leadership position in the sciences and ensure a steady supply of skilled scientists entering the workforce. Two, advocates for diversity argue that a diverse workgroup outperforms a homogenous one, with collective wisdom exceeding the sum of its parts [*Page*, 2007], the case for diversity driven by increased scientific excellence rather than simply as an indicator of the larger U.S. population.

6.7. Creating a Culture of Inclusiveness

With the creation of the OAAD in 2008, there have been concerted efforts to create a culture of inclusiveness via an ongoing process of institutional transformation. *The focal point of these efforts has been the institution and its policies and procedures, and not any specific group or individual.* This is because many of the causes of underrepresentation are systemic, arising due to structural problems. The purpose behind creating an inclusive institution is not simply to afford legal protection to underrepresented groups but to create an enabling environment for individuals to realize their fullest potential and capabilities, and to be able to participate in the activities of the institution as valuable members of the LDEO community. The focus has been on increased institutional accountability and responsibility, with intervention occurring at multiple levels of activity within the institution, and new policies and procedures being devised wherever it was considered necessary. Five broad areas of intervention were identified:
- Search committees
- Family leave policies
- Institutional awareness and outreach
- Advancing junior staff
- Visibility of women and minorities

6.7.1. Search Committees

Search committees play a crucial role in determining the demographic composition of new hires; awareness of diversity issues on the part of search committee members is crucial in promoting diverse searches. If most of the applicants are white male, it comes as no surprise if the selectee is also white male. Starting in 2009, new guidelines and procedures for search committees were put in place to increase diversity among searches for scientific staff. These guidelines were based on findings from social science research and recommendations from ADVANCE, and included information on best practices, subconscious bias, composition of search committees, targeting diverse venues for posting jobs, and the inclusion of OAAD in all searches, with full access to applicant information, search committee composition, and demographics of the applicant pool. These new guidelines led to a dramatic increase in the diversity (i.e., women and/or nonwhite applicants) of the applicant pool.[3] Over the period 2009–2010, 8 out of 10 searches had an applicant pool that comprised at least 70% women and/or minorities, with all searches showing at least 50% diversity. For 2011–2012, every search yielded an applicant pool that was at least 50% female and/or minority. This is in contrast with the period 2007–2009, where the majority of searches had applicant pools with less than 50% women and/or minorities, with some searches having mostly white male applicants.

The increased diversity in the applicant pool led to an increase in diverse hires. Over the period 2009–2012, new hires were approximately 50% white male, with the other 50% female or nonwhite. These search guidelines have now become the default search procedures at LDEO, leading to the expectation that such diversity in the applicant pool, with a concomitant increase in the hiring of women and/or minorities, will be sustained into the future, gradually diversifying the entire scientific staff.

6.7.2. Family Leave Policies

In July 2010, the Lamont Research Professor (LRP) track was implemented at LDEO, replacing the former Doherty Research Scientist track. Some policies for this track were developed after taking into consideration the results of the 2005 ADVANCE survey, especially with respect to institutional support and family leave policies. LDEO operates in a predominantly soft money environment where scientists typically raise funds from federal/external agencies to support their research projects. Typically, parental leave benefits are paid for by the source of the individual's salary; that is, for full-time faculty, the university provides the individual's salary, and would therefore bear the costs of any maternity/parental leave benefits. In the case of a soft money environment this becomes challenging since it is an external agency that is the source of the individual's salary, with the provision of maternity/parental leave benefits often depending on the policies of the external agency.

In such a situation, taking time off for familial responsibilities hinders career advancement, especially for junior women. The LRP track offers maternity/

---

[3]Despite this increase, there has been little increase in the percentage of underrepresented minorities (Blacks, Native Americans, and Hispanics); a large portion of the above increase was from the increase of women and foreign-born Asians in the applicant pool.

parental benefits equivalent to what tenure-track faculty at other Columbia departments receive, with the cost of these benefits being borne by LDEO. In addition, for individuals with small children, the LRP track also offers the option of working part-time while still retaining full-time status (therefore retaining university benefits such as health insurance and housing). Besides these, the LRP track offers stop-the-clock provisions for promotions for up to two times per individual.[4]

These policies are especially important since research [see *Goulden et al.*, 2009] has shown that women tend to bear a disproportionate impact of familial responsibilities. In addition, the biggest leak in the academic pipeline occurs during the postdoctoral years, a pattern confirmed at LDEO. Without institutional support to accommodate these needs, women are less likely to advance in their careers at the same rate as their male counterparts.

It is pertinent to note that after ADVANCE provided funds to women scientists for back-up care, Columbia has now institutionalized back-up care, with the Office of Work Life providing assistance to all Columbia officers who need back-up care for family members. A lactation room was also set up at *Lamont* in [2009], the first on the Lamont campus. The Lamont day care center, located close to the Lamont campus, has also benefited individuals with young children.

6.7.3. Institutional Awareness and Outreach

Diversity and inclusiveness within an institution are linked to factors pertaining to academic affairs such as salary structures, appointments and promotions, institutional governance, and awareness on how decisions impacting the greater body are made. Greater awareness of the workings of the institution enables greater transparency and accountability. Since 2008 there have been sustained efforts to spread awareness about diversity issues and their close connection with academic affairs and institutional governance. The following information has been disseminated using institutional data in order to promote awareness, and to increase institutional accountability:

- Detailed and anonymized salary comparisons of LDEO scientists at different ranks and titles, these salary comparisons being based on race and gender
- The demographic composition of LDEO scientists by race and gender
- The demographic composition of invited seminar speakers by race and gender (LDEO holds several seminars for which it invites speakers who are considered experts in the field; more often than not, these speakers are male, though in recent years there has been an increase in the number of female speakers)

---

[4]Columbia offers stop-the-clock provisions, but the challenges of the soft money environment at LDEO often made it previously difficult for LDEO scientists to use these provisions.

- A summarized version of the LDEO bylaws in an easy-to-understand format, making it easier to understand the workings of the LDEO leadership. This includes information on promotions, appointments, and policy making. The LDEO Directorate is currently working on simplifying the bylaws and promoting greater inclusiveness by redefining voter eligibility requirements among the scientific staff.
- Information on subconscious bias and stereotype threat using social science research. This is disseminated to the Lamont community periodically to maintain a constant awareness on the issue.

6.7.4. Advancing Junior Staff

An integral component of creating a more inclusive culture is to offer support to junior scientists towards their career advancement. Given that the junior ranks at LDEO are significantly more diverse than the senior ranks both in race and gender (postdoctoral scientists are approximately 50% women and 30% non-white), creating an inclusive environment will most likely increase retention, leading to increased diversity across the institution as these junior scientists get promoted to more senior ranks. *One of the key efforts at LDEO in this context has been fostering the creation of a postdoctoral community.* At the time that OAAD was created in 2008, one-on-one confidential meetings with postdoctoral scientists revealed that a large number of them felt isolated from the larger Lamont community, and some did not know any other postdoctoral scientists at LDEO. Based on feedback received from these individuals, the LDEO Directorate took the following steps to create a sense of community among the postdoctoral scientists and to foster a culture of inclusiveness and support:

- *Postdoctoral Luncheon with the Director:* Once a semester, postdoctoral scientists have lunch with the director of LDEO in an informal setting, where they share their thoughts and concerns about their experiences at LDEO. Feedback received during these sessions has laid the groundwork for initiatives that were since developed for the advancement of postdoctoral scientists and junior staff. Postdoctoral scientists indicated that just knowing there were other postdoctoral scientists who faced similar problems and challenges was reassuring, as was being able to chat informally with the director on the issues that they faced. They also indicated that they were overall very happy with their scientific research at LDEO, though guidance on career advancement varied from mentor to mentor.
- *Postdoctoral Mentoring Plan:* Based on discussions with postdoctoral scientists at the Postdoctoral Luncheon with the Director sessions, along with the NSF requirement that proposals requesting funding for postdoctoral scientists must show how the postdoctoral scientist will be mentored, in the fall of 2010 LDEO implemented an institution-wide Postdoctoral Mentoring Plan. Mentors are expected to provide guidance on career paths, advancement

within LDEO, and also help integrate postdoctoral scientists into the scientific community both internally and externally. This initiative has been well received by the postdoctoral scientists.

- **Small Grants Initiative:** Small grants (approximately $500) are awarded to postdoctoral scientists to attend workshops and conferences on career advancement. Upon returning from these conferences/workshops, postdoctoral scientists are expected to give a presentation to other postdoctoral scientists at LDEO to share what they learned, and to encourage other postdoctoral scientists to attend such conferences/workshops. Such presentations are usually done over a lunchtime meeting sponsored by the LDEO Directorate, and they have been well received by the postdoctoral scientists, not just for the information learned about the conference/workshop, but also for fostering a sense of community.
- **Creation of a Postdoctoral Mailing List:** A postdoctoral mailing list is maintained by the LDEO Directorate and is the primary means of communicating with postdoctoral scientists on an LDEO-wide basis (individual divisions have their own mailing lists for their scientific staff but none geared specifically to postdoctoral scientists). Communications include soliciting feedback on any issue of concern, introducing new postdoctoral scientists, and planning workshops and other events involving postdoctoral participation. The mailing list also serves as a community bulletin of upcoming events and announcements.
- **Postdoctoral Symposium:** In April 2012, LDEO held its first-ever Postdoctoral Symposium. The goal of the event was to provide postdoctoral scientists an opportunity to showcase their research in front of the entire Lamont community, and for the Lamont community to learn more about the scientific research that our postdoctoral scientists are engaged in. This event was extremely well received by the entire Lamont community, junior and senior alike, and is expected to become an annual event.
- **Paid Parental Leave:** Starting in March 2013, postdoctoral scientists are entitled to paid maternity leave, with the salary for the leave period (usually 6–8 weeks) being borne by the LDEO Directorate. In addition, postdoctoral scientists (male or female) may also avail of Columbia University's unpaid childcare leave for up to one year to care for a newborn or newly adopted child.

In addition to the above, the following are aimed at helping all junior scientists:

- **Research Life Series:** This is a series of in-house seminars and workshops typically held over the summer, aimed at providing career guidance to junior scientists at LDEO. These sessions encompass a wide range of areas that are especially relevant to junior scientists, such as sessions on grant management, proposal writing, PI responsibilities, promotions and career advancement within LDEO, postdoctoral mentoring, and research ethics and responsible conduct.

- *Workshops:* In addition to the in-house sessions offered by the Research Life Series, external presenters are invited to conduct workshops on topics relevant to the junior staff, including postdoctoral scholars. Topics have included "Navigating the Federal Funding System," "How to Write Effective NSF Proposals," "Talking to the Media," and "Careers in Science Policy."

6.7.5. Visibility of Women and Minorities

Current research indicates that low visibility and recognition of women and racial minorities compared to their male colleagues contributes to leaks in the pipeline and women not attaining senior positions. Some of LDEO's efforts at increasing the visibility have been the following:

- *Marie Tharp Fellowship:* Started during ADVANCE and named after Marie Tharp, a pioneer of modern oceanography who did groundbreaking work on mapping the ocean floor, this prestigious fellowship brings women scientists to Columbia to collaborate with Lamont scientists. In 2010 this award was institutionalized within the LDEO Directorate with support from the Earth Institute. In addition, as word of this fellowship has spread, the quality of the applicants has been increasingly outstanding, including a Minister for the Environment and a former member of the Intergovernmental Panel for Climate Change. The expectation is that these fellows will forge long-lasting ties with LDEO researchers, and will also serve as role models for junior women scientists.
- *Diversity Seminar Series:* Hosted by the LDEO Directorate, this is aimed at exploring the causes and consequences of diversity in multiple domains, and disseminating research findings on the impact of diversity on scholarly excellence, given the past research that has shown that greater educational benefits are associated with more diverse academic communities [*Page*, 2007]. Invited speakers have been eminent scholars and leaders in their respective fields and have included figures such as Claude Steele, former Columbia Provost, an African American and eminent scholar in the field of stereotype threat and subconscious bias.
- *Women in Science Networking Event:* Networking plays an important role in keeping women in the pipeline, and a lack of role models has been cited in the literature as another reason why women leak out of the academic pipeline. In 2010 the LDEO Directorate sponsored a networking event for women scientists, attended by approximately 60 women scientists from 14 institutions in the northeast region of the U.S. The goal of this event was to connect junior women scientists at LDEO with senior women scientists both from Columbia and outside, and for these junior women to hear insights and advice from the senior women on advancing their careers in a male-dominated field. This event was extremely well received and an event summary was published in *EOS* (the weekly newspaper of AGU). It is expected that more such events will be hosted.

- *Excellence in Mentoring Award:* This award recognizes the importance of quality mentoring, which benefits the institution as a whole. Mentoring has also been identified as a key activity that can contribute to promoting diversity and inclusiveness in the sciences. Messages have been sent out to the Lamont community about the importance of recognizing the contributions of our female senior scientists along with those of their male colleagues. Nominations from the Lamont community for this award have included both male and female mentors; the presence of female role models will likely influence more women to pursue science.

6.8. Change in LDEO Gender Demographics 2005–2012

With respect to gender, the most immediate and visible change has been the increase in the number of women across all ranks (junior and senior) as shown in Figure 6.1. This is true for the scientist ranks (Lamont research professor and research scientist) as well as the scientific support ranks (staff associates). This change is especially prominent in the case of Lamont assistant research professors as shown in Figure 6.2, with the percent of women more than doubling from 18% in 2005 to 39% in 2012. At the senior level, however, the scientific staff is predominantly male, increasing from approximately 15% women to 20% women over the period 2005–2012, suggesting that we have a long way to go before achieving gender parity at senior ranks.

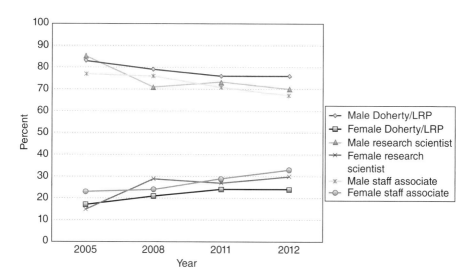

Figure 6.1 LDEO scientific staff by gender, 2005–2012. For color detail, please see color plate section.

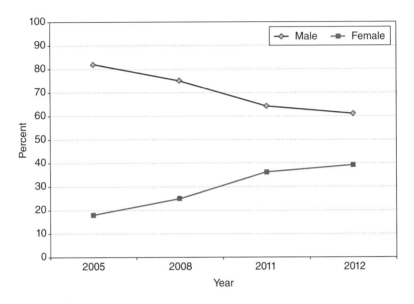

Figure 6.2 Doherty/Lamont assistant research professors by gender, 2005–2012. For color detail, please see color plate section.

6.9. Understanding the Institutional Transformation Process at LDEO

The main focus of LDEO's institutional transformation efforts has been to bring about a systemic change with respect to policies, procedures, and decision-making processes as they pertain to diversity. *At the heart of this systemic change is the inextricable linking of academic affairs and diversity issues,* with these two being so closely integrated that diversity considerations are now routinely brought into any discussion of academic affairs, be it appointments and promotions or salary structures.

Some key features of LDEO's institutional transformation experience have been the following:

- The target of these efforts has been the institution and its workings, not any group or individual. Efforts to promote diversity have *not* been done at the expense of the majority, thereby avoiding a perceived trade-off between diversity and academic excellence.
- The goal of these efforts has been to create *a culture of inclusiveness* and improve the *quality of the work environment*, reflecting the notion that a more inclusive and improved work environment will attract the most outstanding talent to LDEO.
- The Office of Academic Affairs and Diversity plays a hybrid role, integrating diversity concerns into academic decision-making processes, keeping diversity issues on the agenda of all academic decision-making discussions, and

participating at multiple levels of decision making within LDEO, from junior-most to senior-most.
- Efforts have been made to spread greater awareness on social science research pertaining to diversity, with increased institutional mindfulness about the criteria and processes for academic decision making.
- Long-term sustainability has been a key consideration in the development of new initiatives.
- Engaging and involving LDEO scientists in these efforts has been key, especially through their roles as committee members on a wide and diverse range of committees, from award selections and search committees to academic processes. These include both junior and senior scientists across multiple departments.

6.10. Conclusion

The results so far of LDEO's institutional transformation experience have been extremely encouraging. Using social science research as a guide, combined with institutional data, efforts to promote awareness on the importance and relevance of diversity issues have been made at various levels across the institution. The creation of the Office of Academic Affairs and Diversity within the LDEO Directorate, and its role as an *"organizational catalyst,"* being placed at the convergence of academic decision-making processes and institutional activity pertaining to diversity, has been key in promoting inclusiveness and addressing many of the concerns identified in the ADVANCE survey. These activities, focusing on transforming the institutional culture, have sent out a positive message to the LDEO community about evolving institutional priorities and the awareness of the need to diversify our scientific staff.

As is often the case with efforts aimed at transforming the institutional culture, perceived credibility plays a role in influencing their ultimate success. With the ownership of these efforts resting with the LDEO Directorate, the perceived credibility is strong and the general feeling is that the institution is moving in a positive direction. The creation of the Lamont research professorships has been especially well received and is expected to attract outstanding and diverse talent to LDEO. Given that many of the issues and challenges discussed in this chapter are not unique to LDEO but are common to other scientific institutions and departments across the country, the LDEO experience could serve as a model for peer institutions facing similar problems.

REFERENCES

Fisher vs. University of Texas, U.S. Sup. Ct., Docket No. 11-345 (2011).
Goulden, M., K. Frasch, and M.A. Mason (2009), *Staying Competitive: Patching America's Leaky Pipeline in the Sciences*, University of California–Transaction ID

#1MW29424KJ1561044 rkeley; Berkeley Center on Health, Economic & Family Security; and the Center for American Progress.

Grutter vs. Bollinger, 539 U.S. 306 (2003).

Hill, C., C. Corbett, and A. St. Rose (2010), *Why So Few? Women in Science, Technology, Engineering and Mathematics*, AAUW.

National Science Foundation (2008), Division of Science Resource Statistics.

National Science Foundation (2009), Division of Science Resource Statistics.

Nelson, D. (2007), *A National Analysis of Minorities in Science & Engineering Faculties at Research Universities*, available at http://faculty-staff.ou.edu/N/Donna.J.Nelson-1/diversity/Faculty_Tables_FY07/07Report.pdf.

Page, S.E. (2007), *The Difference: How the Power of Diversity Creates Better Groups, Firms, Schools, and Societies*, Princeton University Press, Princeton, NJ.

Sturm, S. (2006), The Architecture of Inclusion: Advancing Workplace Equity in Higher Education, *Harvard Journal of Law & Gender*, 29(2).

U.S. Census Bureau (2011), *Statistical Abstract of the United States*.

7

DUAL CAREER, FLEXIBLE FACULTY

Mary Anne Holmes

Department of Earth and Atmospheric Sciences, University of Nebraska–Lincoln, Lincoln, Nebraska

ABSTRACT

Accommodating dual career couples in academia goes hand-in-hand with creating multiple types of academic positions: non-tenure-leading positions such as professor of practice and research professors, and tenure-leading positions such as split, shared, and temporary part-time. Finding a position for a partner is facilitated by having a campus point person who contacts short-list candidates before they interview and provides a discrete person to whom candidates can reveal personal information. The point person then alerts the appropriate dean and department chair. Academic institutions demonstrate commitment to this crucial component of faculty work-life balance by providing financial support to departments that hire qualified partners, such as three years of bridge funding. Assistance for non-academic partners involves a different skill set: employees with ties to the non-academic, local community. The Dual Career Network and the Higher Education Recruitment Consortium help academic institutions find positions for dual career partners. More than half of men STEM faculty and over 80% of women STEM faculty are in STEM dual career relationships; these proportions are not likely to decrease in the foreseeable future [*Schiebinger et al.*, 2008].

7.1. A Woman's Issue?

When W. J. Lewis became chair of the Math Department at the University of Nebraska–Lincoln, he wanted a way to launch his department into the national spotlight quickly. Having watched his department lose excellent women

Women in the Geosciences: Practical, Positive Practices Toward Parity, Special Publications 70.
First Edition. Edited by Mary Anne Holmes, Suzanne OConnell, and Kuheli Dutt.
© 2015 American Geophysical Union. Published 2015 by John Wiley & Sons, Inc.

candidates for faculty positions when he could not find placements for their husbands, he decided to tackle dual career hiring. His department grew from a single woman on the faculty in the mid-1980s to 10 in 2012, with more than half hired with their partners. At the same time, the department's production of women with advanced degrees rose from 0 to 13 doctoral recipients and 49 master's in the 1990s. For this achievement, the Math Department was awarded the Presidential Awards for Excellence in Science, Mathematics, and Engineering Mentoring in 1998. Hiring outstanding women with an outstanding partner clearly had an impact on hiring and retention of women faculty for this department. Is this the direction for future faculty?

Dual career couples are increasing. In the U.S., 36% of academics are married to an academic partner [*Schiebinger et al.*, 2008]. The rate is much higher for scientists: 54% of men and 83% of women scientists have an academic, scientist partner. Most of these dual career scientist couples are in a similar field, a phenomenon known as disciplinary endogamy [*Schiebinger et al.*, 2008], a type of homophily, or "birds of a feather flock together" phenomenon [e.g., *McPherson et al.*, 2001]. With the high rate of endogamy among women scientists, it seems difficult to hire women scientists without finding a place for their partners, as W. J. Lewis learned, but with more than half of men scientists partnered with another scientist, we see a new faculty phenomenon, not simply a woman faculty phenomenon. In one year across STEM departments at UNL, 40% of the short-listed candidates for 25 positions were dual career. The need for finding a job for a partner is especially critical in college towns where there are few options for nonacademic employment.

7.2. The "Problem" of Dual Career

We would like to propose some terminology for these sometimes delicate issues. Physicists call dual career the "two-body problem" [*Wolf-Wendel et al.*, 2004], but we prefer the term "dual career opportunity." The couple who are the dual career, the "two-body," would rather not be perceived as a problem. Their situation is becoming the new normal. In addition, the partner of the primary hire is often called a "trailing spouse." We prefer the term "opportunity partner." Examples abound of the partner outperforming the primary hire or performing just as well, but the stigma of being a "trailing spouse" is a hard one to live down.

Why is dual career considered a problem? Typically, because after the laborious task of wading through applications and determining a short list of applicants for a faculty position, of spending the time and effort and cost to bring the candidates to campus, of surviving the heated faculty discussion on who to make the offer to, the desired candidate pops up and says: "Oh, by the way, might there be anything for my partner?" Offers have short shelf lives, and highly qualified STEM women are bound to have multiple offers. How does a department chair scramble to accommodate the partner and not lose the candidate? The process presents difficulty when both partners fit into the same department, but negotiating across a college or with a different college across campus can be nearly impossible without personal connections or an *institutional structure* in place that supports dual career hires.

Dual career couples can be a great strength for the department. If both halves are satisfied with their situation, they are more likely to stay and be productive, positive contributors to the department and institution. The institution can hire exceptional faculty who might not otherwise have considered applying, and the couple can obtain satisfying careers for both partners [*Snow*, 2010; *McNeil and Sher*, 2001]. Academia has historically paid less attention to its workers' private lives than industry has, in part because the academy sees a faculty position as so attractive that there is no need to address personal concerns and because there are typically many more applicants than there are positions [*Wolf-Wendel et al.*, 2004]. This historic lack of attention to work-life satisfaction in academia is surprising, considering tenure is a lifetime commitment between the institution and the faculty member [*Wolf-Wendel et al.*, 2004].

7.3. Strategies to Address Dual Career

Dual career couples come in at least two flavors: the partner is or aspires to becoming an academic, or the partner is not an academic. Finding a job for the partner differs by this variable. Academic partners seek a position in the institution or at a nearby institution, possibly a tenure-leading position. If not academic, the partner may have a professional degree or not. Either way, a nonacademic job requires that the institution use a different set of tools to assist in placing the partner. We address these situations separately.

7.3.1. Nonacademic Partners

Finding a nonacademic job requires connections within the local community via the Chamber of Commerce and other local resources. Many institutions have an office with staff who build these connections. Many of these offices are members of the Higher Education Dual Career Network (HEDCN; see Resources below). The HEDCN offers best practices for working with the partner. Best practices include making expectations clear to the partner, knowing institutional policies for placement in staff positions, and effective marketing strategies. Some university dual career offices offer partners professional development skills such as resume writing. Most have a time limit for working with the partner on job placement, usually one year.

The Higher Education Recruitment Consortium (HERC; see Resources below) offers a portal where academic and staff positions are posted. Academic institutions pay a fee to be a member of HERC; job seekers use the job listings on the website for free. The job listings have dual career capability, so that a dual career couple can look for places that offer two appropriate positions. Job seekers can determine how far apart the two jobs can be while searching the job listings. HERC works best in higher density areas where there are several academic institutions in close proximity.

Tech Valley Connect is a private spinoff of an ADVANCE program (Settle In) in the Troy, New York, area (see Resources below). It offers support for families when one or both partners gets a job in the area, helping new hires find schools, daycare, eldercare, dual career positions, and other personal needs. Local businesses contribute to Tech Valley Connect's operating expenses because they gain access to highly qualified people who are moving to the area because their partners got a job there, and so are likely to stay, making the investment in new personnel worthwhile.

7.3.2. Academic Partners

For an academic dual career hire to be successful, both partners must be sufficiently qualified and accomplished that each will be successful independently [*Snow*, 2010]. The process for hire and the expectations of the couple must be explicitly laid out: independence in research (the couple's responsibility), independence in evaluation (the institution's responsibility, particularly at the department level). Certain aspects require an up-front negotiation. For example, how will the couple demonstrate responsibility for research activities if they are research partners [*Snow*, 2010]? Other aspects require long-term processes, such as developing a culture at the institution and within the department that dual career couples are an expected part of our professional lives.

Providing a variety of positions for the academic partner improves the chances of success for all concerned. Non-tenure leading positions established at many academic institutions are listed in Table 7.1 [*Gappa et al.*, 2007; *Lubchenco and Menge*, 1993; *de Wet and de Wet*, 1997; *Goldberg and Sakai*, 1993].

As examples of how split positions work, Jane Lubchenco, who served as administrator of NOAA from 2009–2013, and her husband, Bruce Menge, asked for and received a single position that they split 50:50 at the University of Oregon [*Lubchenco and Menge*, 1993]. Each had left tenure-track positions: she from Harvard and he from the University of Massachusetts–Boston. The position worked well for them when they were starting and raising a young family. The positions were gradually increased to two full-time positions. Carol and Andrew de Wet of Franklin and Marshall University have written extensively of the benefits of split or shared positions for families with young children [e.g., *de Wet and de Wet*, 1997). It is important for the couple to ensure that their position really is part-time, with expectations spelled out clearly for both sides for evaluations [*Monks*, 2009].

Table 7.1 Examples of non-traditional academic positions to increase flexibility for partner hires.

Position	Tenure-Leading?	Funding Stream	Job Expectation
Professor of practice (assistant, associate, full ranks)	No	Institution	Teaching > scholarship
Research professor (assistant, associate, full ranks)	No	Predominantly soft money (i.e., through grants and contracts)	Scholarship > Teaching
Lecturer	No	Institution	Teaching
Shared positions	Yes	Institution ± external sources	Partners share one position; each is evaluated separately. If one fails the tenure bid, the other partner may or may not get the whole position.
Split positions	Yes	Institution + external sources	Partners split one to two positions, e.g., 1.5 positions. Each is evaluated separately. If one partner fails the tenure bid, the other partner may or may not get the whole position. Each person gets a full vote in faculty meetings.
Part-time	Yes	Institution ± external sources	Position may be *temporarily* or *permanently* part-time, with position split into teaching, research, and service loads as negotiated by department and job applicant.

7.3.3. Institutional Mechanism for Academic Dual Career Couples

The University of Nebraska–Lincoln (UNL) has developed a proactive process to address dual career partners through our ADVANCE Institutional Transformation award from the National Science Foundation [*Holmes*, 2012]. Other institutions have a similar process, such as the University of Virginia and Penn State, and so it should be transferrable to other institutions. It addresses the time factor by notifying the candidate of the dual career program once they are short listed and before the interview. It provides the candidate and partner a safe place to disclose this information (the ADVANCE-Nebraska office), which also serves as a relay office to alert all necessary parties. The extra time allows us to have a plan in place to bring the partner to campus to interview as soon as the offer for the primary hire is put together, assuming the partner is qualified for an academic position.

At UNL, all short-listed candidates (the applicants who are invited to campus to interview for a faculty position) are approved through a central administrative office, Equity, Access and Diversity (EAD). The EAD office sends a copy of the approved short list to the ADVANCE-Nebraska office. We send a letter to each short-listed candidate informing them of our dual career program before they come to campus.

The letter informs the candidate that if they have a dual career partner who qualifies for a faculty position, the partner should send a letter of interest, C.V., and statement of research and teaching interests to the ADVANCE office. The director of the ADVANCE program notifies the department chair and the college dean of the potential dual career hire. If the partner fits best in a different department or college, the director relays the information to the appropriate dean for possible forwarding to the department of interest. The major players omitted from this information flow are members of the search committee. This precludes their inappropriately weighing partner status as a criterion for selection of the candidate with the best potential (see chapter 3, discussion of "implicit assumptions"). We are human, after all, and the urge to speculate on the candidate's partner status and how that will impact likelihood of accepting an offer is pretty hard to resist. But dual career potential has little to do with the primary candidate's qualifications for the position.

If the candidate with an academic partner is ultimately selected for a faculty position and the partner's target college and department are interested, the partner is flown to Lincoln for an interview within a week or two of the primary partner's selection. If the interview is successful and the target department faculty vote to make an offer to the partner, two offers go forward.

The ADVANCE grant supplies one fourth of the salary for the partner for up to three years, with the department, college, and provost's office supplying one fourth each as well. The Office of Research pitches in by providing their typical proportion of startup funds for the partner. The three-year bridge funding buys time for all the administrators to "shuffle the chess pieces" (await a retirement or new line) and find a permanent funding stream for the partner. When the grant ends, the provost's (Senior Vice Chancellor for Academic Affairs) office will pick up one half of the bridge funding.

We anticipated up to eight dual career hires over the life of the five-year grant. By the end of year three, we hired ten partners. We had minor challenges: for example, not all partners find that they want to stay in academia. In addition, partners do split up. Overall, we have had made more success with these formal procedures than with prior informal attempts to manage last-minute requests for partner accommodation. The program fills a need and works, in large part because department chairs agree to cooperate: if we accept a partner now, another department will help us in the future.

Department interest and cooperation are essential for a dual career program to be successful. Faculty fear giving up a position in an area of their choosing to hire someone whose specialty is not quite what they want. This can pose an insurmountable barrier to hiring the partner. But we have seen situations where the unexpected specialty becomes a vital new area of excellence in the department, so we suggest that faculty approach the dual career potential with an open mind. Administrators might consider rewarding cooperative departments whenever this is possible by not "mortgaging" a hoped-for position.

7.4. For Dual Career Couples

When is the right time to disclose the need for a job for a professional partner? Reading the above gives the impression that telling the department chair and/or search committee member sooner rather than later allows them time to put together a potential position for the partner. But from your point of view, unless there is some sort of dual career program in place that has notified you before the interview, it is in your best interest to disclose nothing until the offer is made. None of us wants to hear the conversation that a search committee might have: "Let's make the offer to him/her, we'll get a two-fer," or conversely, "We can't make an offer to him/her, we'll never find a position for the partner!" When an actual job offer emerges, you hold the strongest position for negotiation that you will ever have [*Goldberg and Sakai*, 1993].

Be realistic. Small institutions cannot create new faculty lines from nothing. Larger institutions frequently experience budget shortfalls and may not be able to create new faculty lines. Be realistic about the abilities and potential for each person in this dyad. Rarely, a partner who cannot get a tenure-leading line at first can perform brilliantly (i.e., external funding for research and/or scholarship with plenty of publishing) and will earn one eventually. At the least, it can put the two of you in a good position to seek dual employment elsewhere.

RESOURCES

Dual Career Network: https://sites.google.com/site/dualcareer/index

Higher Education Recruitment Consortium: http://www.hercjobs.org/

SERC offers resources and case studies: http://serc.carleton.edu/NAGTWorkshops/careerprep/jobsearch/dualcareer.html

Tech Valley Connect: http://www.techvalleyconnect.com/; http://www.elsevierfoundation.org/new-scholars/stories/video-phd-move.asp

University of Illinois Urbana–Champaign spells out dual career procedures; not the onus of contacting a second department is on the chair of the primary hire department: http://www.provost.illinois.edu/communication/08/Communication_No_8_0112.pdf

REFERENCES

de Wet, Carol B., and P. Andrew (1997), Sharing academic careers: An alternative for pre-tenure and young family dual-career faculty couples, *Journal of Women and Minorities in Science and Engineering*, *3*(3), 203–212.

Gappa, J. M., A. E. Austin, and A. G. Trice (2007), *Rethinking Faculty Work: Higher Education's Strategic Imperative*, Jossey-Bass.

Goldberg, D., and A. K. Sakai (1993), Career options for dual-career couples: Results of an ESA survey on soft money research positions and shared positions, *Bulletin of the Ecological Society of America*, *74*(2), 146–152.

Holmes, M. A. (2012), Working together, *Nature*, *489*, 327–328.

Lubchenco, J., and B. A. Menge (1993), Split positions can provide a sane career track: A personal account, *Bioscience*, *43*(4), 243–248.

McNeil, L., and M. Sher (2001), *Dual-science-career couples: Survey results.* Available at http://www.physics.wm.edu/~sher/survey.html.

McPherson, M., L. Smith-Lovin, L., and J. M. Cook (2001), Birds of a feather: Homophily in social networks, *Annual Review of Sociology*, 415–444.

Monks, J. (2009), Who are the part-time faculty? *Academe.* Available at http://www.aaup.org/article/who-are-part-time-faculty.

Schiebinger, L. L., A. D. Henderson, and S. K. Gilmartin (2008), *Dual-career academic couples: What universities need to know*, Stanford University.

Snow, J. T. (2010), *Working with dual career couples: A 10-year perspective.* Available at http://www.docstoc.com/docs/23661837/Working-With-Dual-Career-Couples-A-Ten-Year-Perspective

Wolf-Wendel, L., S. B. Twombly, and S. Rice (2004), *The Two-Body Problem: Dual-Career-Couple Hiring Practices in Higher Education*, Johns Hopkins University Press.

8

LACTATION IN THE ACADEMY: ACCOMMODATING BREASTFEEDING SCIENTISTS

Suzanne OConnell

Department of Earth and Environmental Sciences, Wesleyan University, Middletown, Connecticut

ABSTRACT

Breastfeeding is a public health issue with a large economic value. Lactation is an important component of work–life balance and family friendly accommodations, especially in male-dominated science, technology, engineering, and mathematics (STEM) fields. Women's primary childbearing years commonly overlap with graduate school, pretenure, and even posttenure employment. Balancing the challenges of work and a new infant are an equity issue for women in the competitive academic environment. A healthy, breast-fed baby helps a mother to remain viable in the academic workforce. University administrators and department chairs should make access to lactation facilities as seamless as possible.

8.1. Why for the Nation?

Breastfeeding is a public health issue with a large economic value (Box 8.1). Among many other benefits, research has shown that breastfed babies are healthier and their mothers are therefore less likely to need to stay home to take care of them and miss work [*AAP*, 1997, 2005]. The importance of breastfeeding was published by the American Academy of Pediatrics (AAP) in 1997 and more strongly reiterated in a revised version published in 2005. This importance was recognized by Congress when, on March 23, 2010, it passed the "Affordable Care Act," commonly known as Obamacare. This legislation requires employers to "provide reasonable break time for an employee to express breast milk for her nursing child for one year after the child's birth each time such employee has need to express the milk. Employers are

---

Women in the Geosciences: Practical, Positive Practices Toward Parity, Special Publications 70.
First Edition. Edited by Mary Anne Holmes, Suzanne OConnell, and Kuheli Dutt.
© 2015 American Geophysical Union. Published 2015 by John Wiley & Sons, Inc.

Box 8.1   Economic and social benefits of breastfeeding (estimated dollar amounts are for different years, from *Silver*, 2010, accessed 2013).

- Parental absenteeism is three times lower
- Infant benefits
  - Reduces infant deaths
  - Less susceptible to acute infectious diseases
  - Fewer infections
  - Lower risk of chronic diseases (e.g., asthma, diabetes, leukemia)
- Maternal benefits
  - Decreases a mother's chances of contracting ovarian cancer
  - Decreases a mother's and daughter's likelihood of breast cancer
  - Decreases postpartum bleeding
  - Faster return to prepregnancy weight
- Lower insurance costs
  - Insurers pay at least $3.6 billion each year to treat diseases and conditions preventable by breastfeeding
  - Health care services attributable to formula feeding cost at least one health maintenance organization $300 more than did services for a breastfed infant

---

also required to provide a place, other than a bathroom, that is shielded from view and free from intrusion from coworkers and the public, which may be used by an employee to express breast milk." This mandate was further supported in February 2011, when the Internal Revenue Service amended its policy, allowing breast pumps and related equipment to be considered medical devices. This allows these items to be purchased with funds from a pretax Health Flexible Spending Account.

8.2. Why in the Academy?

Lactation is an important component of work-life balance and family-friendly accommodations, especially in male-dominated STEM fields. Women's primary childbearing years commonly overlap with graduate school, pretenure, and even posttenure employment. Balancing the challenges of work and a new infant are an equity issue for women in the competitive academic environment. A healthy, breastfed baby helps a mother to remain viable in the academic workforce. University administrators and department chairs should make access to lactation facilities as seamless as possible.

Although most faculty have their own office, and many may feel comfortable nursing or pumping in that space, there are reasons that it might not be conducive to breastfeeding. Offices are a workspace and some women may not be able to relax enough to be able to nurse. Some offices have large windows, allowing office activities to be visible to the campus. Protocol in other departments is that a closed door means the faculty member is not present, and even a locked door can be opened without knocking, exposing the nursing or pumping professor. "College and University Lactation Programs: Some Additional Considerations," by *Barbara Silver* [2010], provides Internet links to lactation programs at 11 public and private research universities.

8.3. What Is Needed?

Each mother's needs vary; therefore, lactation programs need to be flexible. There are three basic components to a successful lactation program: time, space, and support.

1. **Time.** Most women need to express milk every 3 to 4 hours and it can take 20 to 30 minutes. More time is needed to get to and from the lactation room.
2. **Space.** A lactation room should contain a comfortable chair for the nursing mother to sit in and a flat surface to place a breast pump upon. There needs to be a way to keep other people from interrupting privacy, such as a lock on the door and/or a sign that says "occupied" to let people know the room is in use. While not required by law, employers might also consider providing an electrical outlet, a breast pump, a sink (hand and breast equipment washing), a small refrigerator for storing breast milk, and an interior that creates the relaxed environment that nursing moms need to let down their milk and experience a productive, pumping session.
3. **Support.** Expressing milk is easier and quicker if the mother is relaxed, so it may take less time as she becomes used to the process. Support from the university and department includes acknowledging the importance of breastfeeding and providing facilities and policies that encourage this practice.

The American Academy of Pediatrics recommends breastfeeding for the baby's entire first year. Not expressing milk can have detrimental health effects for the mother, such as mastitis (painful swelling and inflammation of the breasts), and infections. Not expressing milk can also reduce milk productivity, requiring milk substitutes for the infant with the complications of illness requiring the mother to be absent from work to care for the sick child.

8.4. Professional Travel

Nursing mothers are part of the work force. They may be visitors to campus for conferences and talks. Identifying a lactation room in or near science buildings will decrease the complications of such travel. Most professional societies now provide lactation rooms at conferences. In a recent discussion, two mothers compared the positive changes between a bare, isolated, curtained room at an American Geophysical Union conference more than five years ago with the more recent, cozier room filled with many nursing and pumping mothers.

RESOURCES

American Academy of Pediatrics (AAP) (1997), Breastfeeding and the use of human milk, *Pediatrics*, *100*, 1035–1039. doi:10.1542/peds.100.6.1035

American Academy of Pediatrics (AAP) (2005), Breastfeeding and the use of human milk, *Pediatrics*, *115*, 496–506. doi:10.1542/peds.2004–2491

United States Department of Labor, *Break Time for Nursing Mothers*, http://www.dol.gov/whd/nursingmothers/. Last accessed May 7, 2013.

About.com, *Are Breast Pumps Tax Deductible?* http://usgovinfo.about.com/od/healthcare/a/Breast-Pumps-Tax-Deductible.htm. Last accessed May 7, 2013.

Kocieniewski, D. (2011), *Breast-Feeding Supplies Win Tax Breaks From IRS*, http://www.nytimes.com/2011/02/11/business/11breast.html?_r=4&

Silver, B. (2010), College and University Lactation Programs: Some Additional Considerations, available at http://www.uri.edu/worklife/family/family%20pics-docs/LactationPrograms%20FINAL.pdf. Last accessed July 30, 2013.

Work and Pump: Support and Information for Mothers Committed to Providing the Best Nutrition for Their Babies, http://www.workandpump.com/. Accessed July 31, 2013.

Workplace Accommodations to Support and Protect Breastfeeding (2010), http://www.usbreastfeeding.org/Portals/0/Publications/Workplace-Background-2010-USBC.pdf.

Working Moms: Tales and Vents about Pumping Breast Milk, http://workingmoms.about.com/u/ua/todaysworkingmoms/PumpingBreastMilkUA.htm.

SECTION III.B: INTERACTIONAL AND INDIVIDUAL STRATEGIES

9

IMPLICIT ASSUMPTION: WHAT IT IS, HOW TO REDUCE ITS IMPACT

Mary Anne Holmes

Department of Earth and Atmospheric Sciences, University of Nebraska–Lincoln, Lincoln, Nebraska

ABSTRACT

Our implicit attitudes influence how we evaluate people for jobs, for admission to graduate school, and for awarding fellowships, scholarships, and professional awards without our being aware of them. Implicit associations that form implicit attitudes develop by repeated contact with a given phenomenon. For example, when every nurse we've ever seen is a woman, we mentally picture a woman whenever we hear the word "nurse." Similarly, most scientists portrayed in the media are men, so most people think of a man whenever they hear the word "scientist." We can minimize the impact of our implicit attitudes on how we evaluate applications by learning about the existence of implicit attitudes, defining the criteria for evaluation before we look at applications, and taking the time to evaluate each applicant based on the predefined criteria. Institutions can assist in minimizing the impact of negative implicit attitudes by promoting consistent evaluation procedures and by holding search, admission, and award nomination committees accountable for having their applicant pools and short lists match the demographics of the available pools.

Ben Barres, a Stanford neurobiologist, gave a seminar at an Ivy League school and afterwards heard a faculty member say of his performance: "Ben Barres gave a great seminar today, but then his work is much better than his sister's." But Barres has no sister in this field. He transitioned from female to male. His performance that day compared the current Ben to that of the woman he used to be [*Barres*, 2006]. Did Ben's scientific abilities improve when he became male? He thinks not.

Women in the Geosciences: Practical, Positive Practices Toward Parity, Special Publications 70.
First Edition. Edited by Mary Anne Holmes, Suzanne OConnell, and Kuheli Dutt.
© 2015 American Geophysical Union. Published 2015 by John Wiley & Sons, Inc.

He suggests that societal assumptions, "implicit attitudes" about who makes the best scientist, explain this comment better.

When we hear the word *scientist* on the radio or TV, what do we picture? Most people around the world think of a man in a white lab coat, and in the U.S., a white man in a white lab coat, just as kids visiting Fermilab sketch *before* they tour the labs (Figure 9.1; http://ed.fnal.gov/projects/scientists/). The original study of children's drawings of scientists, now routinely replicated at Fermilab, revealed six consistent components: a lab coat, glasses, facial hair, symbols of research such as lab equipment, symbols of knowledge such as books, technology (often rockets), and captions depicting the scientist crying, "Eureka!" or speaking in formulas [*Chambers*, 1983]. In addition, there are often sinister images, such as smoking beakers (Figure 9.1, "Before"). Only 28 of 4,807 images analyzed depicted women scientists.

Such a mental image of a person in a particular role is an example of a *schema*, a mental (cognitive) framework that helps us organize the world and use cognitive shortcuts to navigate our way through each day. Schemas become stereotypes when we associate qualities or traits with an entire group. Schemas evolve through our associations over time [*Ridgeway*, 2006]. For example, if every nurse we see is a woman, when someone says "nurse," our mental image is apt to be a woman.

After touring Fermilab and being exposed to the variety of scientists who work at the labs, kids draw a variety of people as scientists (Figure 9.1, "After").

Before   After

Figure 9.1 Children who visit Fermilab are asked to sketch and describe their concept of a scientist before and after they visit the labs and meet the scientists working in them. The "before" sketch is often a white man in a white lab coat. The "after" sketches are more diverse. See http://ed.fnal.gov/projects/scientists/amy.html. For color detail, please see color plate section.

Their description of a "scientist" changes as well, from descriptors such as "annoying, crazy" to "normal, just people with interesting jobs."

Where do these schemas of what a scientist looks like come from? Most likely they are adopted from the portrayals of "mad scientists" in the media [e.g., *Shibeci and Sorensen*, 1983]. Few kids see images that contradict this idea unless they have an opportunity such as visiting the Fermilab or having different kinds of scientists regularly visit their classrooms [*Flick*, 1990; *Finson*, 2010; *Finson et al.*, 2010).

What about scientists: what do scientists think a "scientist" looks like? Do we think of a variety of people when someone says "scientist"? Research in psychology, cognition, and sociology in the last three decades indicates that when scientists hear the word, we also think of the most common type of scientist that we see and interact with every day. This matters because our schema of who is a "scientist" affects who we think should be our next colleague, the next nominee for an important award, the next department chair, named professor, or chair of a powerful committee. Schemas operate below the level of consciousness, and so are unconscious or implicit (as opposed to explicit). Unconscious or implicit attitudes can conflict with our conscious beliefs [*Banaji et al.*, 2003]. Naturally we think we are fair and ethical judges of applications for a new faculty member or awardee. But are we? The research suggests that we operate under the "illusion of objectivity" [*Armor*, 1999]. When implicit attitudes negatively impact an evaluation or judgment outcome, they have morphed into *implicit biases*.

9.1. What Are Implicit Attitudes?

Experiments originally intended to test Freud's theories of the conscious and the unconscious mind demonstrate the influence of internal, unconscious images on our behavior [*Greenwald and Krieger*, 2006]. While Freudian concepts such as the superego and the id were largely abandoned in the twentieth century due to a lack of research-based evidence to support them, the "unconscious" influence on cognition (mental processes) did survive testing: unconscious, implicit memories and experiences do influence how we feel and behave [*Greenwald and Banaji*, 1995; *Greenwald and Krieger*, 2006]. Implicit cognition does not arise from conscious, intentional control. As *Greenwald and Banaji* [1995] explain it, "An implicit [assumption] is the introspectively unidentified (or inaccurately identified) trace of past experience that mediates [behavior or judgment]."

Early research on unconscious influences focused on implicit memories and how they affect the choices people make. People were exposed to some phenomenon, and later they behaved in a way that indicated influence by memory of that exposure. Test participants, however, did not remember the actual exposure, or remembered it imperfectly [*Greenwald and Krieger*, 2006]. For example, in early experiments, people were shown a few letters of a word and later asked to provide a complete word with these letters [*Greenwald and Banaji*, 1995]. One group was shown a list of words prior to seeing the letters. This group used the words from the list they'd seen to complete the words; the control group used different words. When asked to recall the actual list of words, the group shown the list could not do so. Something beneath conscious control influenced word recall: a memory of the list but not an explicit, conscious memory.

Similarly, if we hold an unexamined mental image of who is a scientist, that implicit image will affect how we review applications for the position of a scientist. This will affect who we encourage to become a scientist, who we prefer when we review a pile of applications, and what colleagues we think of when it's time to nominate someone for a prestigious award [*Holmes et al.*, 2011]. Additional experiments to those begun in Greenwald's lab verify the impact of prior experience and acculturation on our value judgments as well, and the results of these experiments bear directly on the success of recruitment, promotion, and retention of women and people of color in STEM. A few studies are mentioned here; these provide further references for readers who wish to delve more deeply into this literature.

9.2. How Do Implicit Attitudes Affect Evaluations?

Faculty propensity to hire a male for a managerial position was demonstrated when 127 faculty from research-intensive universities in biology, chemistry, and physics departments were asked to rank applications from male or female students for a laboratory manager position [*Moss-Racusin et al.*, 2012]. The "applications" were actually a single made-up application, with a male name on one copy and a female name on the other. Both men and women found the male applicant significantly more competent and hirable than the female applicant, and both men and women faculty proposed to offer the male student a higher starting salary. Although the nonexistent applicants had identical resumes, both men and women faculty *implicitly* assumed the male candidate would be preferable in a lab manager position, that is, they did not consciously think, "This job is better held by a man." This study illustrates two important points about implicit bias: (1) we implicitly think men are more preferable for certain types of jobs (and women are implicitly thought more preferable for a different set of jobs, usually lower paying) and (2) both men and women hold the same implicit biases. In this study, women faculty offered the lowest salary for the female applicant: $25,000 per year versus $27,111 offered by men faculty. Men faculty offered the highest to the male applicant: $30,520 versus $29,333 offered by women faculty. Implicit assumptions and the biases that derive from them arise from the shared culture we grow up in and are not intrinsic to one gender or race.

In another example of how we infer competence and fit of an applicant for a position from the very names of applicants, psychology faculty were asked to evaluate a real psychologist's curriculum vitae (CV), but half received the CV with the male name "Brian" and the other half received the CV with the female name "Karen" [*Steinpreis et al.*, 1999]. Both male and female faculty were 50% more likely to prefer hiring "Brian" over "Karen." Name may suggest race of a candidate: in another study, identical resumes sent to employers in the Chicago and Boston areas yielded 50% more callbacks for the fictional "Greg" over "Jamal" [*Bertrand and Mullainathan*, 2003]. Other tip-offs in resumes can trigger our implicit biases, such as marital and parenthood status. Marital status [*Jordan and Zitek*, 2012] and parenthood [*Hodges and Budig*, 2010] implicitly signal stability for men but a lack of commitment for women. In a study by *Correll* and her coworkers [2007], employers preferred to hire nonmothers over mothers because, as study subjects said, nonmothers are "more committed to the workplace." These studies suggest that both men and women see men as more appropriate for the position of "scientist." There is evidence that in situations where negative stereotypes are evoked, both majority and nonmajority people see people of color as less fit for the job [*Jost and Banaji*, 2011]. We are not bad people, but we can have unexamined habits that disadvantage the underrepresented without our intending this to happen.

Implicit bias influences the types of letters of recommendation that are written for graduate school applicants and for job applicants. A detailed examination of over 300 letters of recommendation for student applicants to medical schools revealed that both men and women write longer letters of recommendation for men [*Trix and Psenka*, 2003]. In this study, letters written for men applicants had more superlative language and more direct references to the applicant's CV, whereas letters written for women applicants had more "doubt-raisers" such as "It appears that her health and personal life are stable" and irrelevant comments, such as "She is quite close to my wife" and "It appears that her health and personal life are stable." Looking at the words in letters of recommendation, an analysis of 624 letters for 194 people who applied for eight faculty positions open from 1998 to 2006 in a psychology department at a Southern U.S. university revealed that letters written for men had more "agentic" terms such as *assertive, confident, aggressive, ambitious, dominant, forceful, independent, daring, outspoken,* and *intellectual* [*Madera et al.,* 2009]. Letters written for women contained more "communal" terms such as *helpful, kind, sympathetic, sensitive, nurturing, agreeable, interpersonal, warm, caring,* and *tactful.* It is nice to be described as "helpful," but further results indicate that agentic terms get you the job [*Phelan et al.*, 2008]. In a study of 866 letters of recommendation for a biochemistry faculty position, more "standout words" were used for male candidates [*Schmader et al.*, 2007]. Again, both men and women write similar letters of recommendation and show the same implicit attitudes.

Letters of recommendation may be heavily weighted when we are sifting through a pile of applications for graduate school or the next faculty position or the next recipient of a local or national award. If the letters are slightly biased, does it make a difference in the ultimate outcome? *Martell and others* [1996] performed a computer simulation of an eight-tiered workplace, (i.e., with seven chances for promotion). With equal numbers of men and women in the lowest level and no difference between male and female evaluation scores, promotion through the system produced equal numbers of men and women at the highest level. But with equal numbers of men and women at the lowest level, when women were given a 1% lower evaluation score, their numbers diminished until only 35% of the highest level positions were occupied by women. A 5% difference yielded so few women, that the simulation had to be run with a higher proportion of women at the entry level. Beginning at 60% of the workforce, a 5% disadvantage yielded less than 30% women at the highest level in the simulated organization.

So, yes, it makes a difference in outcomes. Every step that involves evaluation that activates our implicit assumptions about who our next colleague should be and how a scientist looks imparts a small disadvantage to the recipient of implicit bias. "Mountains are made of molehills," as *Valian* [2005] noted.

Focusing on the criteria for evaluation before we see who is in the applicant pool helps us choose who is really the best candidate and not the nearly best one. The costs to a university for a failed search or for hiring the wrong person can be significant: advertising the position, travel expenses for several interviews, and the cost of faculty time spent on application review and interviews, generates an estimated cost of $10,000–$20,000 at University of Nebraska–Lincoln. For a failed tenure bid, the costs are obviously greater, including not only the search costs but the startup costs and the human cost: the sense of failure by the candidate and by the department members, the graduate students left to finish up as best they can. Hiring the right person is fraught with peril. If we skip over a great person because of an accumulation of disadvantage, our institution and our department lose in a big way (see chapter 1).

9.3. Minimizing the Impact of Implicit Bias

Strategies to reduce the impact of implicit bias on our evaluation processes may address the individual, interactional, or institutional level of human interactions [*Risman*, 2004; see chapter 2]. Our implicit assumptions are malleable: we can reduce them at the individual level with effort and motivation [*Blair*, 2002; *Devine, et al.*, 2002; *Rudman et al.*, 2001]. It appears easier to arrive at better outcomes by addressing institutional policies and procedures, by establishing evaluation procedures that are transparent and consistent, and by holding faculty accountable for outcomes [*Bielby*, 2000; *Dovidio et al.*, 2000; *Green and Kalev*, 2008]. We can effectively change outcomes; it's harder to change people.

9.3.1. Individual Strategies

There are two classes of individuals in an evaluation: the evaluator and the person being evaluated. For evaluators, the first step is **recognizing that implicit bias based on implicit assumptions exists** [*Banaji, et al.*, 2003]. One way to uncover our own biases is to take the Implicit Association Test at http://implicit.harvard.edu or www.tolerance.org/hidden_bias.

Taking more time to thoroughly review applications can minimize implicit bias [*Bertrand and Mullainathan*, 2005]. Time pressure triggers our cognitive shortcuts.

Diversity training is offered for search committees by some institutions and in some corporations, but its effectiveness is uncertain and dependent on trainer and trainee characteristics [e.g., *Holladay and Quinones*, 2005]. This is an intervention that is generated by the institution to act on the individual and is discussed further below in the "Institutional Strategies" section below.

The person being evaluated can benefit from **mentoring programs** (see chapters 11 and 12) that explicitly aid applicants in writing an effective CV and statements of interest. They can benefit from **networking** that raises their profiles and enables them to connect with people who can help advance their career.

9.3.2. Interactional Strategies

Interactional change is the most challenging. Most of these strategies focus on the behavior of the evaluating committee.

The most critical step a selection committee can take to reduce the effect of implicit bias is to **discuss and clarify what criteria** they will use to evaluate the candidates before any application is reviewed [*Bielby*, 2000]. As this procedure should be encouraged by the institution and carried in its memory, it is discussed below in "Institutional Strategies."

It is helpful to **discuss implicit assumptions** and their impact on a committee before any evaluative process begins. Explicit instruction to avoid prejudice can reduce implicit and explicit bias [*Lowery et al.*, 2001]. Many NSF proposal review panels now begin with a discussion of implicit bias, and a final review of proposal rank occurs while asking the question "Does the 'fund' column suggest any bias?" There are a variety of resources to initiate this sometimes difficult conversation, including short videos of search committee behavior (http://www.engr.washington.edu/lead/biasfilm/) and numerous theatre groups that act out the search process (e.g., The Cornell Interactive Theatre Ensemble, The Center for Research on Learning and Teaching Players from the University of Michigan, Mizzou ADVANCE Interactive Theatre). There are numerous resources on the University of Michigan's Committee on Strategies and Tactics for Recruiting to Improve Diversity and Excellence (STRIDE) Web site: http://sitemaker.umich.edu/advance/stride_committee.

The **composition of the evaluative committee** matters. Searches yield more diverse results when someone is paying attention to diversity. Many institutions require that a "minority," a person from some underrepresented group, serve on every evaluation committee. This practice leads to unintended and unfortunate consequences: We assume this person can be an articulate and informed spokesperson for an entire race or gender, when perhaps they cannot. He or she usually has a lower status in the institution (perhaps a graduate student or untenured faculty member) and may fear to speak up on behalf of the underrepresented [*Ridgeway*, 2006]. Having one "minority person" is tokenism and triggers stereotyping for the rest of the committee, who may value such input less [*Craig and Feasel*, 1998]. In addition, this person becomes overloaded with service tasks.

The ADVANCE program at the University of California–Irvine created a unique institutional solution to this dilemma (http://advance.uci.edu/equityadvisors. html): equity advisors are senior faculty who are trained in implicit bias and strategies to reduce its impact. They serve on search committees as the "voice for diversity and excellence." UC-Berkeley has adopted this model (http://diversity. berkeley.edu/graduate/gdp/equity_advisors), as have other ADVANCE institutions (e.g., Wright State, Purdue, Iowa State). In the absence of institutional support for an equity advisor program, one or more majority persons in the department could serve in this capacity as part of their service load. Joan Williams' Gender Bias Learning Project is also a great resource: http://www.genderbiasbingo.com/.

Status differential triggers implicit bias [*Ridgeway*, 2006]. Most promoted and administrative people in the university are white males, persons from the majority group. Consequently, we assume lower status when we see a woman or person of color. More egalitarian workplaces where hierarchy is deemphasized and collaborations are emphasized reduce status differential and implicit bias [*Smith-Doerr*, 2004]. Over the long term, exposing faculty to excellent minority scholars can change implicit assumptions [*Aberson and Haig*, 2007]. Invite women and people of color to speak in your department. Extended visits increase the number of positive interactions that may occur.

9.3.3. Institutional Strategies

Institutions, either at the department level or higher, can have a profound impact on reducing implicit bias in evaluations in two ways: by providing formalized procedures for evaluations that are transparent to all and by holding search committees accountable for conducting unbiased searches [*Bielby*, 2000; *Reskin and McBrier*, 2000].

9.3.3.1. Formalized, transparent procedures

"It happens every time. They always find a reason to not hire the woman candidate. This year, they weighed publications the most and she was two short but had more extramural funding; last year, they weighed external funding the most. And everyone acts like we are being fair every time" (Anonymous, personal communication).

This faculty member's comment is an example of "casuistry," varying the criteria for what makes an acceptable candidate to match predetermined or desired outcomes [*Norton et al.*, 2004]. As discussed as an interactional strategy, we need **transparent, explicit criteria** to evaluate candidates fairly. Evaluative

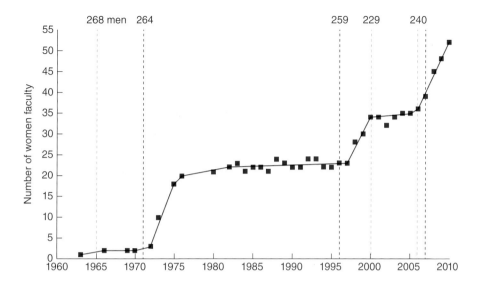

Figure 0.1 Women faculty in the School of Science at MIT (1960–2010). The numbers of women increase only when effort is focused on their recruitment and retention. Between 1970 and 2010, the percentage of women faculty at MIT increased from 8% to 19% [from *Conrad et al.*, 2011].

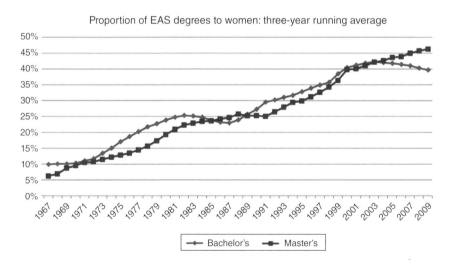

Figure 1.1 Proportion of Bachelor's and Master's degrees in EAS awarded to women. Data from *NSF*, 2013.

Women in the Geosciences: Practical, Positive Practices Toward Parity, Special Publications 70. First Edition. Edited by Mary Anne Holmes, Suzanne OConnell, and Kuheli Dutt. © 2015 American Geophysical Union. Published 2015 by John Wiley & Sons, Inc.

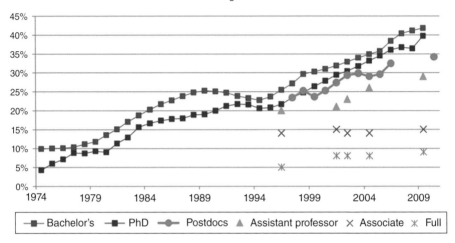

Women in geosciences

— Bachelor's   — PhD   — Postdocs   ▲ Assistant professor   ✕ Associate   ✳ Full

Figure 1.2 Proportion of women at various stages in the geoscience workforce pipeline. Student and post-doc data from *NSF*, 2013. Bachelor's degrees are forwarded by seven years to compare with PhD recipients. Faculty data from *AGI*, 1996–2012, for PhD-granting institutions. Bachelor's and Master's granting institutions have 3–5 higher percentage points of women faculty than doctoral granting institutions.

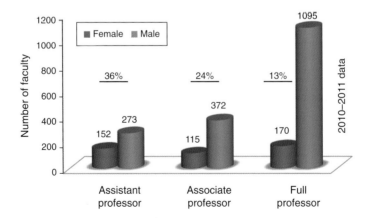

Figure 2.1 Numbers of female and male faculty members by rank at the 106 top-ranked PhD-granting geoscience departments. Data for 2010–2011 academic year.

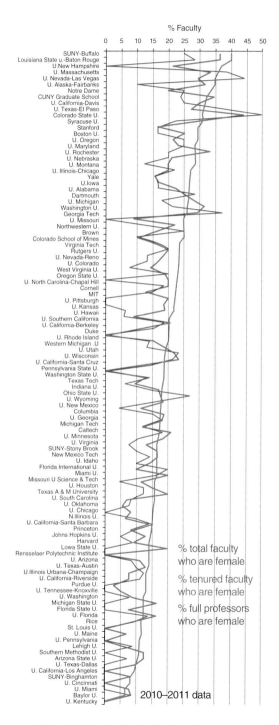

Figure 2.2 Percentage of female faculty by institution for 106 top-ranking PhD-granting geoscience departments in the U.S. Data for 2010–2011 academic year.

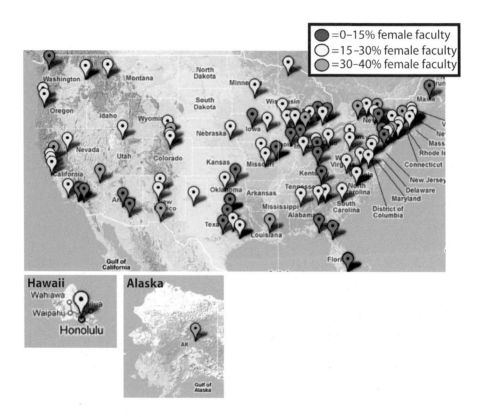

Figure 2.3 Geographic location of top-ranked geoscience departments in the U.S. labeled by color based on percentage of total faculty who are women.

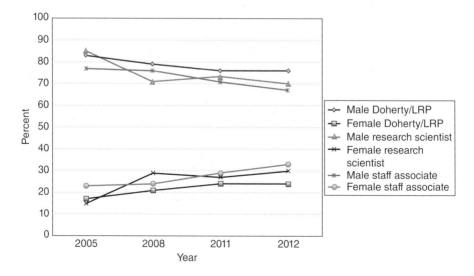

Figure 6.1 LDEO scientific staff by gender, 2005–2012.

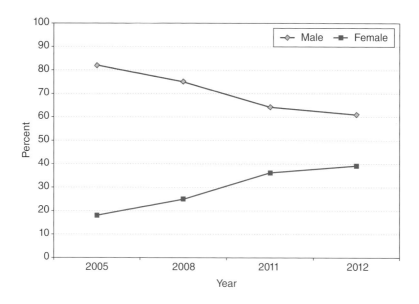

Figure 6.2 Doherty/Lamont assistant research professors by gender, 2005–2012.

Before   After

Figure 9.1 Children who visit Fermilab are asked to sketch and describe their concept of a scientist before and after they visit the labs and meet the scientists working in them. The "before" sketch is often a white man in a white lab coat. The "after" sketches are more diverse. See http://ed.fnal.gov/projects/scientists/amy.html.

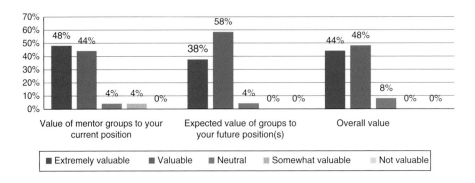

Figure 12.1 Results of MPOWIR's 2012 survey to evaluate the effectiveness of mentor groups.

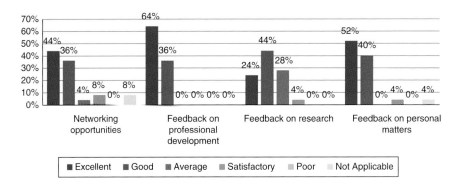

Figure 12.2 Results of MPOWIR's 2012 survey showing value of mentor groups to feedback on professional development and personal matters.

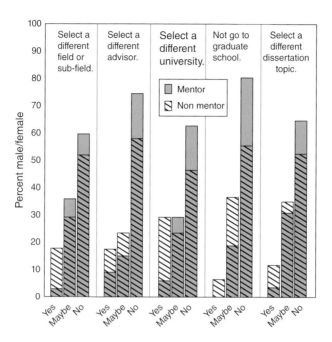

Figure 12.4 Nearly 40% of both male and female respondents do not have a mentor. Whether or not a student has a mentor significantly affects their attitude toward graduate school.

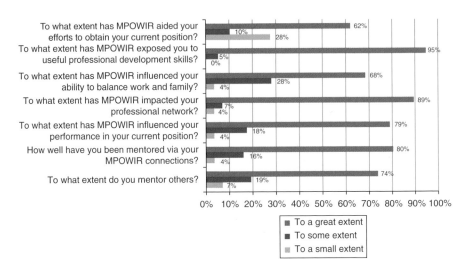

Figure 12.6 Part II of MPOWIR participant survey. Junior women were asked to indicate the overall impact of MPOWIR on their careers.

ASCENT

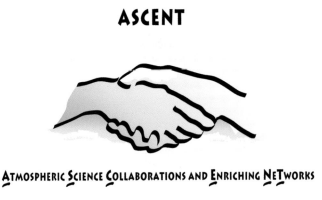

ATMOSPHERIC SCIENCE COLLABORATIONS AND ENRICHING NETWORKS

Figure 13.1 The ASCENT logo created by Lisa Wable, graphic artist at the Desert Research Institute.

committees should agree on the criteria *before* they begin to review applications. This may involve a more formal process of creating a rubric that states what components of the application will be considered and how each component will be rated. Or it could be an agreement of what data will be collected and evaluated. Having a PhD is usually a requirement for an academic position; how important is the institution that granted the degree? If this is important to the committee and the department, then it should be explicitly stated and perhaps a ranking applied: If the degree is from one of these 15 or 20 schools, it gets ranked higher than a degree from this next set of schools. Making the criteria explicit keeps the conversation focused on the candidate's qualifications. Deciding on these criteria before the applications are reviewed helps the committee stay focused on the qualifications and matches a candidate to them, rather than mentally selecting the candidate we like, perhaps for unexamined, implicit reasons, and then arguing for the qualifications that would put that candidate forward.

Work out and state the **ground rules** before the evaluation begins, before the applications are reviewed. Everyone gets a chance to speak and be heard. If graduate students are on the committee, establish beforehand what their role is, particularly whether they can vote or not. Since letters of recommendation are likely to be biased, how will these be weighed in evaluation? Indications that implicit bias affected the letter writer include personal comments, comments on marital status, parenthood, illness, or injuries. The letter should address what is in the CV and some personal attributes such as work ethic. The committee should discuss each candidate's qualifications or lack thereof, and not the candidate's marital status, parental status, or other personal characteristics. "Holes" in CVs must be discounted. Not everyone follows the typical model of undergraduate → graduate → postdoc → faculty position. There are a myriad of good reasons for not doing so. Health issues, health of a family member, or parenthood are a few examples that have no bearing on whether the candidate is qualified or not. Similarly, whether the candidate is in a dual-career partnership must not be weighed for or against a candidate. Institutional procedures to handle dual-career issues should be in place (see chapter 7).

9.3.3.2. Accountability. Holding players within the institution accountable is an effective strategy for minimizing the impact of implicit bias [*Sturm*, 2006]. This may be a formal process, such as having the department chair, college dean, or provost request data from an evaluative committee on the gender and race composition of the applicant pool and on the available pool. Or it may be an informal process in that evaluative committees adopt such a procedure as a self-check. For search committees for new faculty, the available pool includes doctoral candidates and postdocs (obtained from NSF or AGI). The mere act of collecting these data and examining them may be sufficient for a search committee to return to the applicant pool for a reexamination. This is "attention to the demographics of the pool." If the applicant pool does not look like the available pool, why not?

What can the department do differently next time to get a different outcome? Developing a diverse, excellent pool takes time and sustained commitment. *Search is an active verb*; committees should be proactive in finding and recruiting excellent talent to applicant pools (see chapter 10).

9.4. Checklist for Reducing Implicit Bias in Evaluations

• Learn about implicit assumptions; take the Implicit Association Test (http://implicit.harvard.edu/).
• Compose the committee so that someone (not necessarily a person from an underrepresented group) pays attention to and speaks up for qualified candidates who may be overlooked due to implicit bias.
• Write the ad as broadly as possible: women tend to see themselves as less eligible for a position when there are more qualifiers for it.
• Lay ground rules for the committee at the first meeting, before any applications are reviewed; discuss what applicant criteria will be valued and how valued; ensure that everyone gets to speak.
• Establish evaluation criteria and stick to them; omit discussion of personal attributes of the candidate.
• Committee members should take sufficient time to evaluate applications.
• Once the short list is generated, go back through applications and ensure that all candidates were evaluated fairly; was someone missed, perhaps due to implicit bias in the letters of recommendation or other application material?
• For new faculty hires, graduate student, or postdoc visitors, ensure the entire department is aware of appropriate interviewing procedures; know what questions cannot be asked in the interview.

REFERENCES

Aberson, C.L., and S. C. Haag (2007), Contact, perspective taking, and anxiety as predictors of stereotype endorsement, explicit attitudes, and implicit attitudes, *Group Processes & Intergroup Relations*, *10*(2), 179–201. doi: 10.1177/1368430207074726

Armor, D. A. (1999), *The illusion of objectivity: A bias in the perception of freedom from bias*, Doctoral dissertation, University of California–Los Angeles. http://0-search.proquest.com.library.unl.edu/docview/304443547?accountid=8116

Banaji, M. R., M. H. Bazerman, and D. Chugh (2003), How (un) ethical are you? *Harvard Business Review*, *81*(12), 56–65.

Barres, B. A. (2006), Does gender matter? *Nature*, *442*(7099), 133–136. doi:10.1038/442133a

Bertrand, M., and S. Mullainathan (2003), Are Emily and Greg more employable than Lakisha and Jamal? A field experiment on labor market discrimination (No. w9873), National Bureau of Economic Research. http://www.nber.org/papers/w9873

Bertrand, M., and S. Mullainathan (2005), Implicit discrimination, *The American Economic Review*, *95*(2), 94–98. http://www.jstor.org/stable/4132797

Bielby, W. T. (2000), Minimizing workplace gender and racial bias, *Contemporary Sociology*, *29*(1), 120–129. http://www.jstor.org/stable/2654937

Blair, I. V. (2002), The malleability of automatic stereotypes and prejudice, *Personality and Social Psychology Review*, *6*(3), 242–261. doi: 10.1207/S15327957PSPR0603_8

Chambers, D. W. (1983), Stereotype images of the scientist, the draw-a-scientist test, *Science Education*, *67*(2), 255–265. doi: 10.1002/sce.3730670213

Correll, S. J., S. Benard, and I. Paik (2007), Getting a job: Is there a motherhood penalty? *American Journal of Sociology*, *112*(5), 1297–1339.

Craig, K. M., and K. E. Feasel (1998), Do solo arrangements lead to attributions of tokenism? Perceptions of selection criteria and task assignments to race and gender solos, *Journal of Applied Social Psychology*, *28*(19), 1810–1836. doi: 10.1111/j.1559-1816.1998.tb01347.x

Devine, P. G., E. A. Plant, D. M. Amodio, E., Harmon-Jones, and S. L. Vance (2002), The regulation of explicit and implicit race bias: The role of motivations to respond without prejudice, *Journal of Personality and Social Psychology*, *82*(5), 835–848. doi: 10.1037/0022-3514.82.5.835

Dovidio, J. F., K. Kawakami, and S. L. Gaertner (2000), Reducing contemporary prejudice: Combating explicit and implicit bias at the individual and intergroup level. In *Reducing Prejudice and Discrimination: The Claremont Symposium on Applied Social Psychology*, S. Oskamp (ed.), pp. 137–163, Lawrence Erlbaum Associates Publishers, Mahwah, NJ.

Finson, K. D., 2010. Drawing a scientist: What we do and do not know after fifty years of drawings, *School Science and Mathematics*, *102*(7), 335–345. doi: 10.1111/j.1949-8594.2002.tb18217.x

Finson, K. D., J. B. Beaver, and B. L. Cramond, 2010. Development and field test of a checklist for the Draw-A-Scientist Test, *School Science and Mathematics*, *95*(4), 195–205. doi: 10.1111/j.1949-8594.1995.tb15762.x

Flick, L. (1990), Scientist in residence program improving children's image of science and scientists, *School Science and Mathematics*, *90*(3), 204–214. doi: 10.1111/j.1949-8594.1990.tb15536.x

Green, T., and A. Kalev (2008), Discrimination-reducing measures at the relational level, *Hastings Law Journal*, *59*, 1435. Univ. of San Francisco Law Research Paper No. 2010–26. Available at SSRN: http://ssrn.com/abstract=1322504

Greenwald, A. G., and M. R. Banaji (1995), Implicit social cognition: Attitudes, self-esteem, and stereotypes, *Psychological Review*, *102*(1), 4–27. doi: 10.1037/0033-295X.102.1.4

Greenwald, A. G., and L. H. Krieger (2006), Implicit bias: Scientific foundations, *California Law Review*, *94*(4), 945–967. http://www.jstor.org/stable/20439056

Hodges, M. J., and M. J. Budig (2010), Who gets the daddy bonus? Organizational hegemonic masculinity and the impact of fatherhood on earnings, *Gender & Society*, *24*(6), 717–745. doi: 10.1177/0891243210386729

Holladay, C. L., and M. A. Quinones (2005), Reactions to diversity training: An international comparison, *Human Resource Development Quarterly*, *16*(4), 529–545. doi: 10.1002/hrdq.1154

Holmes, M. A., P. Asher, J. Farrington, R. Fine, M. S. Leinen, and P. LeBoy (2011), Does gender bias influence awards given by societies? Eos, *Transactions American Geophysical Union*, *92*(47), 421. doi:10.1029/2011EO470002

Jordan, A. H., and E. M. Zitek (2012), Marital status bias in perceptions of employees, *Basic and Applied Social Psychology*, *34*(5), 474–481. doi: 10.1080/01973533.2012.711687

Jost, J. T., and M. R. Banaji (2011), The role of stereotyping in system-justification and the production of false consciousness, *British Journal of Social Psychology*, *33*(1), 1–27. doi: 10.1111/j.2044-8309.1994.tb01008.x

Lowery, B. S., C. D. Hardin, and S. Sinclair (2001), Social influence effects on automatic racial prejudice, *Journal of Personality and Social Psychology*, *81*(5), 842–855. doi: 10.1037/0022-3514.81.5.842.

Madera, J. M., M. R. Hebl, and R. C. Martin (2009), Gender and letters of recommendation for academia: Agentic and communal differences, *Journal of Applied Psychology*, *94*(6), 1591. doi: 10.1037/a0016539.

Martell, R. F., D. M. Lane, and C. Emrich (1996), Male-female differences: A computer simulation, *American Psychologist*, *51*(2), 157–158. doi: 10.1037/0003-066X.51.2.157

Moss-Racusin, C. A., J. F. Dovidio, V. L. Brescoll, M. J. Graham, and J. Handelsman (2012), Science faculty's subtle gender biases favor male students, *Proc Natl Acad Sciences*, *109*(41), 16474–16479. doi: 10.1073/pnas.1211286109

Norton, M. I., J. A. Vandello, and J. M. Darley (2004), Casuistry and social category bias, *Journal of Personality and Social Psychology*, *87*(6), 817–831. doi: 10.1037/0022-3514.87.6.817

Phelan, J. E., C. A. Moss-Racusin, and L. A. Rudman (2008), Competent yet out in the cold: Shifting criteria for hiring reflect backlash toward agentic women, *Psychology of Women Quarterly*, *32*(4), 406–413. doi: 10.1111/j.1471-6402.2008.00454.x

Reskin, B. F., and D. B. McBrier (2000), Why not ascription? Organizations' employment of male and female managers, *American Sociological Review*, *65*(2), 210–233. http://www.jstor.org/stable/2657438

Ridgeway, C. L. (2006), Linking social structure and interpersonal behavior: A theoretical perspective on cultural schemas and social relations, *Social Psychology Quarterly*, *69*(1), 5–16. doi: 10.1177/019027250606900102

Risman, B. J. (2004), Gender as a social structure: Theory wrestling with activism, *Gender & Society*, *18*(4):429–450. doi: 10.1177/0891243204265349

Rudman, L. A., R. D. Ashmore, and M. L. Gary (2001), "Unlearning" automatic biases: The malleability of implicit prejudice and stereotypes, *Journal of Personality and Social Psychology*, *81*(5), 856–868. doi: 10.1037/0022-3514.81.5.856

Schibeci, R. A., and I. Sorensen (1983), Elementary school children's perceptions of scientists, *School Science and Mathematics*, *83*(1), 14–20. doi: 10.1111/j.1949-8594.1983.tb10087.x

Schmader, T., J. Whitehead, and V. H. Wysocki (2007), A linguistic comparison of letters of recommendation for male and female chemistry and biochemistry job applicants, *Sex Roles*, *57*(7), 509–514. doi: 10.1007/s11199-007-9291-4

Smith-Doerr, L. (2004), *Women's work: Gender equality vs. hierarchy in the life sciences*, Lynne Rienner Publications, Boulder, CO.

Steinpreis, R. E., K. A. Anders, and D. Ritzke (1999), The impact of gender on the review of the curricula vitae of job applicants and tenure candidates: A national empirical study, *Sex Roles*, *41*(7), 509–528. doi: 10.1023/A:1018839203698

Sturm, S. (2006), The architecture of inclusion: Advancing workplace equity in higher education, *Harvard Journal of Law & Gender*, *29*(2), Columbia Public Law Research Paper No. 06-114. Available at SSRN: http://ssrn.com/abstract=901992

Trix, F., and C. Psenka (2003), Exploring the color of glass: Letters of recommendation for female and male medical faculty, *Discourse & Society*, *14*(2), 191–220. doi: 10.1177/0957926503014002277

Valian, V. (2005), Beyond gender schemas: Improving the advancement of women in academia, *Hypatia*, *20*(3), 198–213. doi: 10.1353/nwsa.2004.0041

10

HIRING A DIVERSE FACULTY

Suzanne OConnell[1] and Mary Anne Holmes[2]

[1] *Department of Earth and Environmental Sciences, Wesleyan University, Middletown, Connecticut*
[2] *Department of Earth and Atmospheric Sciences, University of Nebraska–Lincoln, Lincoln, Nebraska*

ABSTRACT

The percentage of women entering tenure-track science faculty positions continues to rise, yet the number of women faculty with the rank of full professor remains small. Faculty tend to be white and in science, technology, engineering, and mathematics (STEM) departments primarily male, while their students reflect the more diverse society we are becoming. This chapter provides information and suggestions on hiring a more diverse faculty. Steps include educating search committees, providing resources for locating and recruiting diverse faculty, wording job descriptions so as to ensure that they appeal to both male and female candidates, promoting awareness on implicit bias and ways to minimize them, and successfully negotiating the hire.

Recruiting and hiring a diverse faculty doesn't happen by accident, but it can happen if it is an institutional goal. The section below provides information and suggestions about ways to help diversity happen (Figure 10.1). It begins with a search committee committed to diversity and excellence.

10.1. Recruitment Before the Formal Search Begins: Insuring a Diverse Applicant Pool

10.1.1. Educating Search Committee

With the passage of the 1964 Civil Rights Bill, discrimination against hiring women and minorities became illegal, but subtle or implicit biases and unconscious associations continue. Considerable research shows that hiring diverse

Women in the Geosciences: Practical, Positive Practices Toward Parity, Special Publications 70.
First Edition. Edited by Mary Anne Holmes, Suzanne OConnell, and Kuheli Dutt.
© 2015 American Geophysical Union. Published 2015 by John Wiley & Sons, Inc.

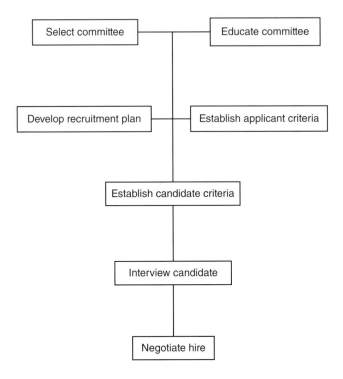

Figure 10.1 Steps to hiring diverse faculty.

faculty is more than a pipeline issue. Search committee members need to be educated about successful approaches for hiring a diverse faculty. It might make sense for the committee as a group to discuss their own approaches to diversity. This discussion could be enhanced by viewing presentations that discuss some of the more subtle aspects of diversity such as Nancy Hopkins's presentation on the "Status of Women in Science and Engineering at MIT" presented as part of MIT's 150th anniversary symposia (http://mit150.mit.edu/symposia/women-of-MIT) or Virginia Valian's interactive tutorials on diversity (http://www.hunter.cuny.edu/gendertutorial/). Professor Hopkins chaired the committee that wrote the 1999 "Report on the Status of Women Faculty in the School of Science at MIT," and her presentation follows the progression of women faculty at MIT. Virginia Valian is the author of *Why So Slow?* Her four tutorials are available as spoken PowerPoint presentations or transcripts, and include an annotated bibliography of resources. These presentations focus on the subtle biases and associations that have prevented and continue to prevent women from being hired in academic positions.

10.1.2. Resources for Locating and Recruiting Diverse Applicants

Key to achieving a diverse applicant pool is to recruit proactively throughout the year and to have broad, open searches rather than focused, narrow searches. Women make up over 35% of geoscience PhD recipients, and although underrepresented minority recipients are not abundant, they do exist and are being produced in increasing numbers (Table 10.1). To insure a diverse applicant pool, search committee members need to actively recruit diverse applicants. One way to do this is to identify research institutions that graduate female and minority PhDs in the search discipline area. If you aren't sure which schools these are, Box 10.1 lists some possible resources.

10.1.3. Job Description for Recruiting Diverse Applicants

We all have unconscious associations, what are sometimes called implicit biases (see chapter 9). However, if they recognize the existence of these biases, search committee members can implement strategies to reduce their impact. All search committee members should take the Harvard implicit test, https://implicit.harvard.edu/implicit/demo/featuredtask.html. This can be a revelatory experience: a disclaimer on the site warns participants, *"If you are unprepared to encounter interpretations that you might find objectionable, please do not proceed further."*

Some words have gender implications, and their use in job advertisements may make a difference in who applies. *Gauchier et al.* [2011] showed that job ads for male-dominated fields contain more words with male stereotypes such as *competitive*, *leader*, and *superior*, and that those for positions dominated by

Table 10.1 Applicant availability. Women and underrepresented minority geoscience doctorate recipients between 2006 and 2010 (from http://www.nsf.gov/statistics/srvydoctorates). % Female includes all females as a *percentage* of total PhD recipients. Ethnic groups include *number* of PhD recipients.

Subject	% Female	Black	Hispanic	Ntv Amer/ Alaska Ntv	Mixed/ Other	Pac Isl/ Ntv Hawaiian	Total White/ Non-Hisp
Atmospheric sci. & meteorology	30	24	n.a.	n.a.	61	n.a.	448
Geological & earth sciences	36	61	141	5	184	n.a.	1677
Ocean/marine sciences	41	8	n.a.	n.a.	62	n.a.	608

n.a. = Not available because the numbers are too low.

Box 10.1   Potential sources of information on recent PhD graduates from underrepresented groups.

The **Committee on Institutional Cooperation (CIC)** is a higher education consortium and includes thirteen primarily Midwestern schools (the Big Ten Conference and the University of Chicago). These institutions grant approximately 15% of U.S. doctoral degrees. Their CIC Doctoral Directory is a searchable listing of doctoral degree recipients from member universities who are members of groups underrepresented in higher education. The directory is designed to increase the visibility of alumni who bring diverse perspectives and experiences to higher education. The directory includes a free online database available to the public. http://www.cic.net/students/doctoral-directory/introduction

Higher Education Recruitment Consortium (HERC) is a nonprofit consortium of more than 600 colleges, universities, hospitals, research labs, government agencies, and related non- and for-profit organizations. They offer a large database of higher education and related jobs that is international in scope. Consortium member institutions share a commitment to hiring the most diverse and talented faculty, staff, and executives. The consortium also offers a comprehensive approach to enabling dual-career couples to find the right jobs within a commutable distance of one another. http://www.hercjobs.org/

The **International Association for Geoscience Diversity** promotes access, accommodation, and inclusion for students and geoscientists with disabilities. Currently the association, which is relatively new (2008), does not maintain a directory of geoscientists with disabilities. As it builds a community, however, the association might become a future resource for finding and hiring candidates. http://www.theiagd.org/

IMDiversity is dedicated to providing career and self-development information to all minorities, specifically African Americans, Asian Americans, Hispanic Americans, Native Americans, and women. It maintains a large database of available jobs, candidate resumes, and information on workplace diversity, although only a small percentage is applicable for faculty jobs. http://www.imdiversity.com

The **Lewis-Stokes Alliance for Minority Participation (LSAMP)** is an NSF program that supports sustained and comprehensive approaches that facilitate achievement of the long-term goal of increasing the number of students who earn doctorates in STEM fields, particularly those from populations underrepresented in STEM fields. The program goals are accomplished through the formation of alliances. This URL lists awardee schools and their awards, http://www.nsf.gov/awardsearch/advancedSearchResult?QueryText=LSAMP&ActiveAwards=true&#results

Table 9 of this **NSF statistics** Web page lists the top 20 PhD granting institutions for minority doctorates in all STEM fields for U.S. citizens and permanent residents. http://www.nsf.gov/statistics/sed/2011/data_table.cfm

Nemnet is a national minority recruitment firm committed to helping schools and organizations in the identification and recruitment of minority candidates. It posts academic jobs on its Web site and gathers vitae from students and professionals of color. Although its focus is precollege, many colleges have used its services (e.g., Cornell University, Florida State University). It also conducts workshops to help organizations recruit minority faculty. http://www.nemnet.com

women contain words with female stereotypes, such as *sympathetic*, *support*, and *interpersonal*. Positions held by roughly equal numbers of males and females did not contain gender bias words. While this may not be such an issue in science, women and minorities still fight the sense of belonging and words are part of that. Possibly such subtle differences contribute to the decrease in women in academic geoscience positions relative to their number as PhD recipients (>35% PhD recipients, ~26% assistant professors) (*Holmes and OConnell*, 2007). One strategy for minimizing this sort of bias is to show some people you would like to attract a draft of the advertisement and see if they find it appealing (Table 10.2).

Women may not apply for top positions. Possibly, this is because they are more likely to deal with problems of self-efficacy, and have seen few people similar to them in such positions [*Murphy et al.*, 2007]. This may make it more difficult for them to imagine that they have the credentials necessary for the job. Members of the search committee should attend professional meetings where presentations and posters will be given and make a point to talk to diverse potential candidates and encourage them to apply. If there is a late-stage diverse graduate student or someone beginning a postdoc, encourage her or him to apply even if her or his dissertation isn't completed. If the person is competitive, it might be worth waiting an extra year for a new faculty member who meets both disciplinary and diversity needs.

While most professionals are likely to have a professional partner, women are still more likely than men to have such a partner and possibly in the same discipline, as found by *Schiebinger* [2008]. This study also found that women consider their partner's status and employability important to their career decisions and refuse job offers if their partner cannot find a satisfactory position.

The department must be prepared to address the two-body opportunity early. *Holmes* [2012] describes an innovative approach developed at the University of Nebraska–Lincoln to being prepared to address dual-career needs. If such accommodations exist at your institution, mention this in the job description.

Table 10.2 Tips on defining and writing the position description from the University of Rhode Island.

What	Why/Research Rationale
Develop broad hiring goals to cast a wide hiring net. Write a clear and specific position description about primary job responsibilities, overview of the department/institution, and commitment to diversity in students, teaching, and research—combined with flexibility in qualifications for the position.	To *avoid weeding out applicants* by sounding austere and inflexible. Your goal is to *attract applicants by making a favorable impression.* Simple linguistic conveniences can help: (1) provide detailed, relevant information both about the posted job and the institution, (2) persuade applicants to generate a favorable impression of what the institution has to offer them, (3) try to sell the department/institution to an applicant rather than attempting to screen them out, and (4) endorse diversity at all levels versus displaying tokenism at lower levels while revealing your motivation for doing so.
Write two position descriptions—what is minimally necessary versus what is desirable. Assess if the "desirable" description will unnecessarily limit the applicant pool.	
Use the words "preferred" versus "required" and "should" instead of "must" to broaden the applicant pool without significantly altering the nature of the position.	
Strategize support of your new hire: cluster hiring, collaborative networks, and facilities.	

(Source: http://www.uri.edu/advance/RecruitTutorial/BeforePosDesc.htm; also see example at http://www.uri.edu/advance/RecruitTutorial/BeforeProactiveLang.html)

10.2. Reviewing Applications and Selecting Candidates for the Short List

10.2.1. Confronting Associations and Biases

The most important step you can take at this point in the process is to have objective criteria that have been established before the applications are reviewed. Continue to learn about bias and unconscious associations. As shown by *Moss-Racusin et al.* [2012], women are as likely as men to harbor gender biases. In their study, both male and female faculty, when hiring a laboratory assistant, considered men more qualified and offered them a higher starting salary. Four other examples of studies showing gender bias in academic science are listed below, but the number of such published studies is much larger.

1. *Wenneås and Wold* [1997] found that a woman applying for a postdoctoral fellowship with the Swedish medical Research Council had to be two-and-a-half times more productive than a man to get the same scientific competence scores by referees. Their study also showed that connections to any of the reviewers, by any of the applicants, increased their competence scores.
2. *Steinpreis et al.* [1999] showed bias in hiring by both male and female psychologists when they were sent identical curriculum vitae for a job applicant distinguished only by the gender of the applicant's name. Curiously, when the participants received male and female versions of curriculum vitae of an early tenure candidate, no gender distinction was made.
3. *Trix and Psenka* [2003] found statistically significant differences in reference letters for male and female applicants. Letters for female applicants were shorter, contained more personal information, praise was more measured, and success was more likely to be attributed to luck. They were more likely to be described as students and teachers rather than researchers.
4. *Budden et al.* [2008] found a 33% increase in papers authored by women in *Behavioral Ecology* when the reviewers were not aware of the authors' gender.

10.2.2. Ways to Minimize Our Biases and Associations

10.2.2.1. Attention and focus. Published studies about the importance of being able to focus on evaluation come from outside the academy where study subjects are more plentiful. *Martell* [1991] found that gender bias of work performance evaluations of male and female police officers was only removed when the evaluator was able to fully focus on the evaluation task. When the evaluator was distracted, male police officers were given higher performance ratings. Similarly, jurors with less time to process information were more likely to respond to racial stereotypes in assigning guilt and the severity of punishment, whereas no stereotype effect was found when jurors' cognitive load (self-paced) was lowered [*Van Knippenberg et al.*, 1999].

Mechanisms need to be in place to require and reward a high level of commitment from search committees. When the University of Michigan implemented procedures outlined by their ADVANCE committee, Strategies and Tactics for Recruiting to Improve Diversity and Excellence (STRIDE), the percentage of women hired in the three major science colleges (Medicine, Engineering, and Natural Sciences) rose from 13% to 29%.

10.2.2.2. Establish evaluation criteria. As stated above, establishing evaluation criteria early is important. The criteria may vary depending upon the needs of the institution. One good place to start designing evaluation criteria is with an assessment developed by the University of Michigan's STRIDE program, (Figure 10.2).

Applicant selection tool

Applicant's name: _____

Please indicate which of the following are true for you (check all that apply):

☐ Read applicant's CV
☐ Read applicant's statements (re research, teaching, etc.)
☐ Read applicant's letters of recommendation
☐ Read applicant's scholarship (indicate what): _____

Please rate the applicant on each of the following:	Excellent	Good	Neutral	Fair	Poor	Unable to judge
Evidence of research productivity						
Potential for scholarly impact / tenurability						
Evidence of strong background in [relevant fields]						
Evidence of [particular] perspective on [particular area]						
Evidence of teaching experience and interest (including grad mentorship)						
Potential to teach courses in core curriculum						
Potential to teach the core curriculum on [particular area] (including creation of new courses)						

Other comments?

_____

Figure 10.2 Example of an application evaluation tool for a junior faculty position developed by the University of Michigan ADVANCE STRIDE program. It can be used as a template and modified. Available at http://sitemaker.umich.edu/advance/good-practices.

10.2.2.3. Creativity. Creativity is also important in recruiting and hiring. Possibly, different hiring departments in a college or division could keep each other informed about the status of their diversity efforts and results of their search committees. Ideally, a department with a long-term interest in diversity could begin courting diverse candidates early in their graduate career.

A nonacademic example of creativity in diversity hiring is Etsy, an e-commerce company. In 2011 with only 6% female engineers, the company made it a priority to hire female engineers but succeeded in only hiring 1 woman out of 20 new hires. In summer 2012, they invested in scholarships for a computer-intensive course for 23 women and hired 5. Other women began applying when they heard about the firm's innovative approach, and now several other companies are using the same investment method [*Kamenetz*, 2012].

10.2.2.4. Joint evaluation. Traditionally, academic candidates' campus interviews are conducted individually. Yet according to a recent experimental study [*Bohnet et al.*, 2012], individual evaluations are much more likely to reinforce implicit gender bias than joint evaluations. This would be a radical change in academic hiring, but it might be necessary if the gender and racial mix of the academy is to be changed.

10.3. Interviewing Candidates / Campus Visit

10.3.1. Campus Visit

As anyone who has been through the interview process knows, it can be a harrowing experience. In today's market, it is likely that all of the interviewing candidates are excellent and well qualified for the position. When this is the case, a campus visit becomes primarily an opportunity for the search committee and department, as well as the candidate, to judge about "the fit" of the candidate into the culture and academic goals of the department. This is subtle and crucial.

Fit, however, does not mean "more of the same." When a department is interviewing a diverse candidate it is important that more than one of "that kind" of diversity be included in the interview pool. This provides more opportunity for diverse candidates to be judged on their merit, rather than as the diversity type.

As with reading applications, search committees should develop a rating system for use with candidate interviews (Figure 10.3). Faculty who are making the decision (voting?) should include information about their preparation for meeting the candidate and the time they spent with a candidate. This can decrease the impact of someone who has a strong opinion but hasn't participated fully in the interview process. When ranking candidates, it might help to open discussion to rank them by different criteria, such as specialty, teaching, scholarship, and pace of productivity.

Candidate evaluation tool

Candidate's name:

Please indicate which of the following are true for you (check all that apply):

☐ Read candidate's CV

☐ Read candidate's scholarship

☐ Read candidate's letters of recommendation

☐ Attended candidate's job talk

☐ Met with candidate

☐ Attended lunch or dinner with candidate

☐ Other (please explain):

Please comment on the candidate's scholarship as reflected in the job talk:

Please comment on the candidate's teaching ability as reflected in the job talk:

Please rate the candidate on each of the following:	Excellent	Good	Neutral	Fair	Poor	Unable to judge
Potential for (Evidence of) scholarly impact						
Potential for (Evidence of) research productivity						
Potential for (Evidence of) research funding						
Potential for (Evidence of) collaboration						
Fit with department's priorities						
Ability to make positive contribution to department's climate						
Potential (Demonstrated ability) to attract and supervise graduate students						
Potential (Demonstrated ability) to teach and supervise undergraduates						
Potential (Demonstrated ability) to be a conscientious university community member						

Other comments?

Figure 10.3 Example of a candidate evaluation tool for a junior faculty position developed by the University of Michigan ADVANCE STRIDE program. It can be used as a template and modified. Original available at http://sitemaker.umich.edu/advance/files/CandidateEvalForm.pdf

10.3.2. Interview Questions

Many interview questions which people with good intentions might consider normal information gathering are in fact illegal. No academic institution wants to confront a lawsuit. The University of Michigan offers a list of topics with examples of

Table 10.3 Examples of legal and illegal topics and questions that can be asked during any employee interview.

Topic	Legal Questions	Discriminatory Questions
Family Status	Do you have any responsibilities that conflict with the job attendance or travel requirements? (Must be asked of all applicants)	Are you married? What is your spouse's name? Do you have any children? Are you pregnant? What are your childcare arrangements?
Race	None	What is your race?
Religion	None You may inquire about availability for weekend work.	What is your religion? Which church do you attend? What are your religious holidays?
Residence	What is your address?	Do you own or rent your house/apartment/condo? Who resides with you?
Sex	None	What is your gender?
Age	If hired, can you offer proof that you are at least 18 years old?	How old are you? What is your birthdate?
Citizenship or Nationality	Can you show proof of your eligibility to work in the U.S.? Are you fluent in any languages other than English? (May ask only as it relates to the job being sought.)	Are you a U.S. citizen? Where were you born?
Disability	Are you able to perform the essential functions of this job with or without reasonable accommodation?	Are you disabled? What is the nature or severity of your disability?

(Source: http://www.hr.umich.edu/empserv/department/empsel/legalchart.html)

legal and illegal ways to ask questions about the candidate, reproduced as Table 10.3. Everyone who will be meeting with the candidate should review this information.

During the interview process for diverse candidates include time for them to meet faculty in other departments who are part of their diversity group. For example, female and minority candidates would probably want to meet other women or minorities to get a sense of the climate and how faculty like them are treated. No one wants to work at an institution where they won't be comfortable or welcomed. Although this might seem inconsistent, the candidate will still want to be evaluated on the basis of their qualifications and this needs to be made clear.

All candidates should learn about family-friendly policies, and search committee members should be aware of institutional policies. These include child care, family leaves for parental care, birth or adoption, and dual-career policies.

10.4. Negotiating the Hire

No department wants an unsuccessful search. Searches are expensive in terms of both time and money. And in today's fiscal climate there is always the possibility that, if hiring is delayed, the faculty position will be rescinded. Make sure that the candidate being interviewed feels welcome and the department is united.

When you make the offer and bring the recruit to campus, make sure he or she is warmly welcomed and that all members of the department are on board about the hire. The chair, or whoever is negotiating the start-up package, should make it clear that the recruit's interests and needs are a top priority. And if the recruit arrives with a partner, make sure the partner is also welcome and shown respect.

REFERENCES

Bohnet, I., A. van Geen, and M. H. Bazerman (2012), When performance trumps gender bias: Joint versus separate evaluation. Retrieved from http://gap.hks.harvard.edu/when-performance-trumps-gender-bias-joint-versus-separate-evaluation.

Budden, A. E., T. Tregenza, L. W. Aarssen, J. Koricheva, R. Leimu, and C. J. Lortie (2008), Double-blind review favours increased representation of female authors, *Trends in Ecology & Evolution* (Personal edition), *23*(1), 4–6.

Gauchier, D., J. Friesen, and A.C. Kay (2011), Evidence that gendered wording in job advertisements exists and sustains gender inequality, *J of Personality and Social Psych*, *10*, 109–128.

Holmes, M. A. (2012), Working together, *Nature*, *489*, 327–328. http://www.nature.com/naturejobs/science/articles/10.1038/nj7415-327a

Holmes, M. A., and OConnell, S. (2007), Why do women remain curiously absent from the ranks of academia? Leaks in the pipeline, *Nature*, *446*, 346.

Kamenetz, A. (2012), How Etsy attracted 500 percent more female engineers, http://www.fastcolabs.com/3005681/how-hack-broken-gender-dynamics-workplace.

Martell, R. F. (1991), Sex bias at work: The effects of attentional and memory demands on performance ratings of men and women, *Journal of Applied Social Psychology*, *21*(23), 1939–1960.

Massachusetts Institute of Technology (1999), A study on the status of women faculty in science at MIT, *MIT Faculty Newsletter*, *11*, 4. http://web.mit.edu/fnl/women/women.html

Moss-Racusin, C. A., J. J. Dovidio, V. L. Brescoll, M. J. Graham, and J. Handelsman (2012), Science faculty's subtle gender biases favor male students, *Proceedings of the National Academy of Sciences*, *109*, 16474–16479.

Murphy, M., C. Steele, and J. Gross (2007), Signaling threat: How situational cues affect women in math, science and engineering settings, *Psychological Science*, *18*, 879–885.

National Science Foundation, National Center for Science and Engineering Statistics (2013). *Women, Minorities, and Persons with Disabilities in Science and Engineering: 2013*, Special Report NSF 13-304, Arlington, VA. Available at http://www.nsf.gov/statistics/wmpd/.

Schiebinger, L. (2008), *Gendered Innovations in Science and Engineering*, Stanford University Press, Stanford, CA.

Steinpreis, R. E., K. A. Anders, and D. Ritzke (1999), The impact of gender on the review of the curricula vitae of job applicants and tenure candidates: A national empirical study, *Sex Roles*, *41*(718), 509–528.

Trix, F., and C. Psenka (2003), Exploring the color of glass: Letters of recommendation for female and male medical faculty, *Discourse and Society*, *14*, 91–220.

Valian, V. (1999), *Why So Slow? The Advancement of Women* MIT Press, Cambridge, MA.

Van Knippenberg, A., A. Dijkersterhuis, and D. Vermeulen (1999), Judgement and memory of a criminal act: The effects of stereotypes and cognitive load, *European Journal of Social Psychology*, *29*(2–3), 191–201.

Wenneås, C., and A. Wold (1997), Nepotism and sexism in peer-review, *Nature*, *387*, 341–343.

11

MULTIPLE AND SEQUENTIAL MENTORING: BUILDING YOUR NEST

Suzanne OConnell

Department of Earth and Environmental Sciences, Wesleyan University, Middletown, Connecticut

ABSTRACT

This chapter looks at the importance of mentoring. After defining what a mentor is, it describes key points pertaining to successful mentoring and offers suggestions for mentors or advisors on how to develop a mentoring plan. It briefly discusses some of the literature on mentoring and discusses activities that are important to both mentors and mentees. Finally, it encourages open and frank communications between the mentor and mentee.

11.1. Introduction

Success in the academy is a combination of many factors. Intelligence and hard work are essential but not sufficient by themselves. Help from mentors and advisors in learning how to navigate the complex corridors of the academy is also fundamental; it is unlikely that someone will master this process unaided. Unfortunately for the outsider, multiple studies have shown that workers in any field tend to mentor and advocate for people who are similar to themselves [e.g., *Chesler and Chesler*, 2002, *McGuoire*, 2002]. To break this pattern, mentors and mentees, students and faculty, insiders and outsiders, chairs and administrators need to examine the importance of passing information between groups and make sure this transmission occurs.

We like to think of mentoring as a process analogous to a bird building a nest. Birds inhabit many different types of environments, just as there are different academic niches. As a result, birds build nests in different locations and with different materials. Although some aspects of nest building may be

---

Women in the Geosciences: Practical, Positive Practices Toward Parity, Special Publications 70. First Edition. Edited by Mary Anne Holmes, Suzanne OConnell, and Kuheli Dutt.
© 2015 American Geophysical Union. Published 2015 by John Wiley & Sons, Inc.

instinctual, there is clear evidence that much of it is also a learned practice [*Walsh et al.*, 2011]. There are many aspects of succeeding in the academy that also need to be learned.

Mentoring is considered so important by the National Science Foundation that postdocs funded through its programs are required to include a mentoring component in their proposal. But what is a mentor? What is his or her function and how does that differ from an advisor? The English language fails us here because it is difficult to distinguish the difference between an advisor and a mentor. Here we consider an advisor to be an assigned position and/or someone providing a single piece of advice. Someone in an assigned position can be a mentor, but doesn't necessarily have to be.

According to the *National Research Council* [1997], "In the broad sense … a mentor is someone who takes a special interest in helping another person develop into a successful professional." This outstanding book describes specific steps for improving mentoring throughout an institution, and between mentors and mentees. It describes specific roles and how to mentor within those roles, as faculty advisor, career advisor, role model, and career consultant. The different roles are by no means mutually exclusive. Most important, however, it says that mentoring is so essential "that it must be embedded in institutional systems of rewards and promotions." Although this statement was published in 1997, we know of no institution where mentoring has yet reached this level of emphasis.

This may be because in what had been a predominantly white male culture, the assimilation proceeded without the need for a formal process of mentoring: it could happen in locations such as the locker room or at a social event, or through visual and verbal clues that might not be picked up by outsiders. These locations, however, might not be as open or comfortable to the new and more diverse members of the academic workforce. In addition, visual and verbal clues that are clear within a dominant culture might not be as clear to someone outside that culture [*Dovidio et al.*, 1988]. With increased diversity, both mentors and mentees may be asked to reexamine their roles in order to enable both of them to develop a relationship which will allow both to prosper, with the mentor taking an active role in enhancing the development and career of the mentee, and the mentee being receptive to advice.

By the time someone finishes graduate school and accepts an academic position, he or she will certainly have had academic advisors, quite likely as an undergraduate and certainly as a graduate student. Most science careers are begun as undergraduates, and at most academic institutions faculty become advisors to students. Usually there isn't much preparation for the role of advising students other than having once been one. Similarly, more seasoned faculty, and especially chairs, may be expected to take on the role of faculty advisor or mentor without much preparation.

11.2. Faculty Chair: Identifying Mentors and Their Roles

Mentoring is at its most basic level a relationship. It can be part of a defined relationship, such as an academic advisor or department chair, or it can grow out of other relationships. Key to the relationship is respect. Box 11.1 offers a list of ways to build respect; the list may also be viewed as containing potential stumbling points for both parties. Although Box 11.1 is presented with a faculty mentor, such as a departmental chair, in mind, it is applicable to any mentoring relationship. As you read through the list, consider how your reactions and comments as mentor or mentee might differ depending upon the gender, race, ethnicity, or economic background of the other person. This is an opportunity to be especially aware of implicit biases, stereotype threat, and imposter syndrome.

In many departments there is no formal mentoring. When this is the case, the chair, or whoever will be writing an annual evaluation, should advise or mentor a

---

Box 11.1   Keys to successful mentoring (adapted from *National Research Council*, 1997).

1. **Take the new faculty member seriously**. A question or problem that seems trivial or irrelevant to you might not be, or it might mask a more serious issue.
2. **Don't dictate answers**. Suggest paths, give the pros and cons of different options, but let the mentee make the final decision. The relationship might benefit from the mentee explaining the reasons for the decision.
3. **Be direct and frank**. This can be uncomfortable.
4. **Belonging**. When most people around you don't look like you, it is easy to assume that you don't belong. Praise is not abundant in the academy, but it can help to mediate harsh proposal reviews or unpleasant student course comments.
5. **Invite other mentors**. Many complex tasks are necessary to develop as a successful academic: writing, speaking, politics, teaching, researching, etc. How does anyone prepare for this? Enlist reinforcements. It might be that the mentee is more comfortable discussing problems with someone who will not be part of his/her promotion and tenure committee.
6. **Meet on neutral ground.** You (the mentor) are the commander of your office and lab. It might not be an ideal place to discuss your mentee's concerns. Select an inexpensive place on or off campus for coffee or a meal, or the library. This allows the mentee to suggest a neutral meeting place where the mentee can afford to offer to "pick up the tab."

new faculty member or find a mentor for this person. Mentoring is a complex job, and it is unlikely that a single person will be knowledgeable about all of the needs of a new faculty member. We discuss the need for multiple mentors in the next section. But if there is an assigned advisor in this formal relationship, he or she should establish specific times to meet and go over progress. This will allow the advisor to notice where additional help is needed and facilitate an introduction to someone who can assist the new faculty member with his or her adjustment.

11.3. New Faculty (Mentee): Examples of Multiple Mentors

As a new faculty member, it is your job to make sure you get the advice you need to succeed. Your career is at stake. Begin with listing what needs to be accomplished professionally during the time available before tenure, promotion to full professor, or whatever you'd like to accomplish after promotion to full. Think about the many facets of professional and nonprofessional life that need mentoring and especially how complicated the decisions became at the end of formal academic training when you began an academic job. What research questions to address? How to set up a lab? What skills will I need? How to make time for writing? If and when to have children? What about soul fulfilling activities? Do I belong in this profession?

Given this complexity, is there one person who can mentor you in all of these facets? Probably not. As discussed by *Sutkowski* [2011] and *Rockquemore* [2011a], different types of mentors are needed. But both of these articles assume you have already established your goals. In fact, by the time you reach this point in your career, you may well have defined professional goals: for example, finish PhD, get a job, write a paper, get tenure, and so on. The other goals, those of work-life balance and soul-fulfilling activities, also need to be considered but may not be as readily defined, will vary considerably, and are not specifically addressed in this document. However, a useful and short book that discusses this is *Minsker*, [2010].

Sutkowski [2011] emphasizes the importance of applying diverse viewpoints to goal setting and suggests the creation of an informal "Kitchen Cabinet" of mentors. His suggested cabinet (Table 11.1) is professionally focused and consists of five members: friend, role model, insider, veteran, and teacher. To his list we would add advocate, someone you can depend upon to take your side and promote you professionally. Cabinet members' perspectives and roles may overlap, but make sure there is someone filling all of the positions that are needed.

Rockquemore takes multiple mentoring further. In a series of articles in *Inside Higher Education* (2011a, b), she takes a comprehensive and detailed view of mentoring, expanding beyond a professional focus. Her approach to mentoring requires the mentee to be proactive, to ask what is needed to succeed, and then to identify people to help to meet those needs. Identifying these people is

Table 11.1 Modified kitchen cabinet after *Sutkowski* [2011].

Title	Characteristics
Friend/ confidant	This friend is one with whom you can share and discuss your professional and personal goals. He/she is probably not in your field, and can give you a broader perspective on your career and life goals.
Role model	This person has the skill set and position you would like to achieve and is someone who can help you acquire that skill set.
Insider	An insider has probably been working at your institution longer than you, understands its inner dynamics, and can help you to become aware of your performance within its larger context. Be careful, however, that this is not a polarizing person or someone whose experience has created bitterness. More than one insider might be useful.
Veteran	This is a more traditional mentor, someone with broad experience in your field who may or may not be at your institution. Establish set times to consult with this person about your progress.
Teacher	Find a person to help you learn the skills you need to progress.
Advocate	Advocates go out of their way to promote and support you professionally, making you aware of opportunities and nominating you for important professional positions and awards, as well as introducing you to key people at your institution and/or in your field.

especially crucial at transitional career stages, such as from graduate school to tenure track, and tenured to full professor.

She suggests creating a chart listing specific needs (Figure 11.1). The mentee is at the center and around her are categories, such as those in the "Kitchen Cabinet" (Table 11.1). Figure 11.1 is an excellent place to start, since there are probably needs that a mentee is not aware of, but there may also be positions listed that you do not need. Underneath the categories are spaces for names. This requires that you put actual names in each of the positions. Even reading and imagining the chart can be intimidating. For example, her ideal intellectual community requires several people who will read the very beginning drafts (0%–25% written) of a paper or proposal. We are not sure how you find such people when time is so scarce for everyone.

We also suggest including a simpler table in your mentor/mentee portfolio, maybe constructed with an advisor and focused on specific tasks. Using both Table 11.1 and Figure 11.1, make a table like Table 11.2, of what type of tasks you are likely to need this year, and possibly a different one looking ahead five years. Then put the name of potential people to fulfill that role. This is best done with someone knowledgeable about your institution. Include in this table how you are going to identify and contact someone to fulfill that role. You do not need someone's permission to assign him or her a role; the person may not even know that

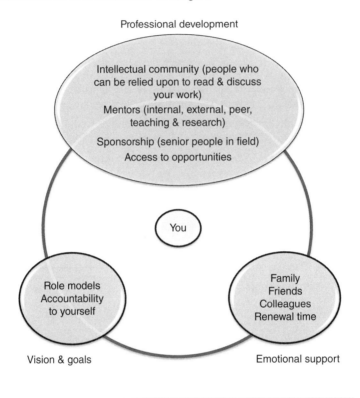

Figure 11.1 Mentoring chart. These positions are not set in stone and a position could be filled by an organization. There may be other people whom you need, if so, add them. There may be people in this chart whom you don't need. Eliminate them. Modified from Rockquemore workshop (www.facultydiversity.org/).

their name appears on your chart or in your table. Similarly, they do not need to be told they are being removed.

Advantages of creating different types of mentoring tables or charts include these:

1. **Continuity.** All of the positions are unlikely to turn over at the same time.
2. **Concreteness.** You need to think concretely about how you are being mentored.
3. **Clarity.** Names on paper make it easier to discuss which needs are being met and which are not, so that you can then find people to complete the chart.

The components, both positions and names, of a mentoring table should be reviewed at least annually, especially as your professional and personal paths progress. Ideally, an institution will have established mentoring mechanisms in

Table 11.2 Example of mentoring task table.

Task	Likely Mentor	How to Contact/ Get Introduced	Comment/ Reminders	Important Dates
Publish XYZ paper this year	Prof. Smith	Advisor will introduce	If 2 weeks go by without any writing progress, see Prof. Smith immediately	Intermediate due dates leading to paper completion
Getting known in field	Prof. Coyne	Send her your C.V. and ask to meet	– Find funds to attend small international meeting – How to connect with NSF program manager?	
Write a final exam	Mr. Jones	He's in teaching and learning center	See at beginning of semester	This week

place. But if it does not, you as the mentee have the most at stake and must take the lead. We believe that awareness and action in this process may contribute towards a healthier departmental culture with benefits for everyone.

11.4. Mentor and Mentee Collaboration

Although everyone is responsible for his or her own professional and personal development, it is possible and even likely that someone new to an academic position, be it graduate student, postdoc, new professor, newly tenured professor, or new chair, may not know how to access mentors or create a mentoring table or chart.

We suggest that the more experienced person, an advisor to a graduate student, department chair to a new professor, or provost to a new chair, take the lead in organizing an advising session that includes mentoring. Show examples of mentoring tables or charts and discuss what needs to be done to succeed in the available time. Considering the investment of time and money involved in recruiting a new faculty member, especially one who has extensive start-up costs, everyone should be behind helping to make this person successful. This mentoring is especially critical for outsiders such as women in science and minorities in any academic field. Imagine how much smoother transitions would be if every new graduate student and new faculty hire were presented with the information about how to get the mentoring needed to succeed and help to plan the next career stage.

At whatever stage, mentors should be aware of common pitfalls, make the mentee aware of them, and provide guidance about how to avoid them.

11.4.1. Three Steps for an Advisor or Mentor

1. **Initiate a meeting**. Before the meeting, both mentor and mentee should review the mentoring information and charts. What are the likely stumbling blocks? For example, a new assistant professor needs to set up a lab, learn to write, and learn to organize and teach a class. At the meeting use Table 11.1 and Figure 11.1 to fill out your own table and a task table (Table 11.2) to determine what advice someone in your position needs to succeed, and to identify specific people who can fill the roles and how those people will be contacted. Include questions such as these: What do I need to accomplish to get tenure? How important are external reviews to the process? If external review is important, are there funds to invite some of these people to campus?
2. **Establish a plan and regular check-in times** to see how the mentee is proceeding, say midsemester, end/beginning of the semester, and midsummer. It might not be possible to proceed on all of the tasks at the same time. Which ones are most important will depend upon the institution. Get help where it is needed: a teaching or writing coach, perhaps, or someone to help with work-life balance.
3. **Emotional support** can come from multiple sources including friends, partners, parents, and siblings. But do not be hesitant to use or suggest professional therapy, religious support, and less common experiences or practices such as meditation, exercise, and nonacademic social groups.

11.4.2. What's Really Important (Mentor and Mentee)?

The importance of the tasks listed below varies considerably by institutional type and even within different departments at the same university. Identify what needs to be done at your school and department by looking at the accomplishments of recently tenured and promoted faculty.

11.4.2.1. Setting up a research lab. Make sure the new hire has the tools needed in start-up funds to succeed. This needs to be negotiated before a new person accepts a position and should be discussed at the time of the job offer. Several papers about negotiating start-up funds are included in the resources section.

11.4.2.2. Publishing. Publishing is the currency of the academy. *Boice* [1990] says, and we agree, that learning to write should be the top priority of a new faculty member. It is the foundation of the nest. This is how someone builds a reputation. Figure out (mentee)/show (mentor) how to make this happen, how to inform the new faulty member about the essential components of a writing practice. Writing is not something to be done at the end of the day, after everything else is done. It must be a priority. Lack of publications is the most common cause of failure to attain tenure. If someone has a strong publication record and does not get tenure, he or she will be in a much better position to find another academic position.

Set an established time to write every week that is as inviolable as teaching a class. Some research suggests that 90-minute segments are ideal, but find out what works for you. Employ an editor to review your material before it gets sent to a journal. If creating figures is a time sink, see if technical illustrators are available to hire. It's possible that students in a technical illustration program are available to hire for not much money.

An important aspect of creating your plan is length of time to publication. Some journals take over a year to accept or reject an article. This means one less year on your tenure or promotion timescale.

If you are having trouble, there are many self-help books on the topic of writing (see Resources), and there are online coaching programs that you can join. Some of these resources are listed in the resource section. Participation in some of these programs can be included in start-up packages.

11.4.2.3. Teaching. At some institutions, the teaching load is light and not very important for career success. At others, good teaching is essential. If you are at an institution that requires good teaching, find out how teaching is evaluated and who does teaching well, then seek their assistance. Teaching can be both a duty and delight, but it is certainly more fulfilling when it is done well.

As with writing, there are many self-help books to learn better teaching techniques. Some of these are also listed in the resources section. In the geosciences, SERC, the Science Education Research Center at Carleton College (http://serc.carleton.edu/), is a tremendous resource for ways to think about teaching, providing class modules, exercises, and syllabi. If you don't find information about a topic that you're looking for, you can probably find other people in the SERC community who are also interested and develop a workshop to cover the topic.

11.5. A Last Word of Caution: Don't Ignore Difficult Topics

As faculty and students become more diverse, the opportunities to interact with and mentor someone who is not like you increase. There are important differences, such race, gender, sexual orientation, disabilities, and economic background. Both mentor and mentee need to be able to openly and frankly discuss uncomfortable topics. Maybe in preparation both parties should read books such as *Can We Talk About Race?* [*Tatum and Perry*, 2007] and *Why So Slow?* [*Valian*, 1998].

REFERENCES

Boice, R. (1990), *Professors as Writers: A Self-Help Guide to Productive Writing*, Forums Press, Stillwater, OK.

Chesler, N.C., and M. A. Chesler (2002), Gender-informed mentoring strategies from women engineering scholars: On establishing a caring community, *Journal of Engineering Education*, *91*(9), 49–455.

Dovidio, J. F., S. L. Ellyson, C. F. Keating, K. Heltman, and C. E. Brown (1988), The relationship of social power to visual displays of dominance between men and women. *Journal of Personality and Social Psychology*, *54*, 233–4242.

McGuoire, G. M. (2002), Gender, race, and the shadow structure: A study of informal networks and inequality in a work organization, *Gender & Society*, *16*, 303.

Minsker, B. (2010), *The Joyful Professor: How to Shift from Surviving to Thriving in the Faculty Life*, Maven Mark Books, Milwaukee, WI.

National Research Council (1997), *Adviser, Teacher, Role Model, Friend: On Being a Mentor to Students in Science and Engineering*, Washington, DC: The National Academies Press. http://www.nap.edu/openbook.php?record_id=5789

Rockquemore, K. A. (2011a), Don't talk about mentoring, *Inside Higher Ed*, http://www.insidehighered.com/advice/mentoring/debut_of_new_column_on_mentoring_in_higher_education_careers.

Rockquemore, K. A. (2011b), Sink or swim, *Inside Higher Ed,* http://www.insidehighered.com/advice/mentoring/essay_on_the_problems_with_the_sink_or_swim_mentality_in_higher_education

Sutkowski (2011), Kitchen cabinet of mentors, *Inside Higher Ed*, http://www.insidehighered.com/advice/2011/07/06/essay_on_the_importance_of_having_multiple_mentors.

Tatum, B., and T. Perry (2007), *Can We Talk About Race? And Other Conversations in an Era of School Resegregation*, Beacon Press, Boston.

Valian, V. (1998), *Why so slow? The advancement of women,* MIT Press, Cambridge, MA.

Walsh, P., M. Hansell, W. D. Borello, and S. D. Healy (2011), Individuality in nest building: Do southern masked weaver (Ploceus velatus) males vary in their nest-building behavior? *Behavioral Processes*, *88*(1), 1–46.

RESOURCES

Mentoring: Articles and Books

Association for Women in Science (1993), *Mentoring Means Future Scientists*, Association for Women in Science, Washington, DC.

Council of Graduate Schools (1990), *Research Student and Supervisor: An Approach to Good Supervisory Practice*, Council of Graduate Schools, Washington, DC.

Council of Graduate Schools (1995), *A Conversation About Mentoring: Trends and Models*, Council of Graduate Schools, Washington, DC.

Fort, C., S. J. Bird, and C. J. Didion (eds.) (1993), *A Hand Up: Women Mentoring Women in Science*, Association for Women in Science, Washington, DC.

Kanigel, R. (1986), *Apprentice to Genius: The Making of a Scientific Dynasty*, Johns Hopkins University Press, Baltimore.

National Research Council (1997), *Adviser, Teacher, Role Model, Friend: On Being a Mentor to Students in Science and Engineering*, The National Academies Press, Washington, DC. http://www.nap.edu/openbook.php?record_id=5789

Olmstead, M. A. (1993), Mentoring new faculty: Advice to department chairs, *CSWP, A Newsletter of the Committee on the Status of Women in Physics*, *13*(1), 811, American Physical Society, Washington, DC.

Roberts, G. C., and R. L. Sprague (1995), To compete or to educate? Mentoring and the research climate, *Professional Ethics Report*, *8*(1), 67, Fall.

Mentoring: Online Resources

Institute for Broadening Participation, Pathways to Science, Mentoring Manual (http://www.pathwaystoscience.org/manual.asp?sort=Overview%20and%20Home)

Mentor Net is an international, nonprofit organization focused on connecting mentors and mentees (www.mentornet.net).

New Faculty (Any Faculty)

Boice, R. (2000), *Advice for New Faculty Members: Nihil Nimus*, Allyn and Bacon (Pearson), Needham Heights, MA [*nihil nimus* means nothing in excess].

Lang, J. M. (2005), *Life on the Tenure: Lessons from the First Year*, Baltimore, The Johns Hopkins University Press.

Rockquemore, K. A., and T. Laszloffy (2008), *The Black Academic's Guide to Winning Tenure—Without Losing Your Soul*, Lynne Rienner Publishers, Boulder, CO.

Toth, E. (1997), *Miss Mentor's Impeccable Guide for Women in Academia*, University of Pennsylvania Press, Philadelphia.

Teaching

Bain, K. (2004), *What the Best College Teachers Do*, Harvard University Press, Cambridge, MA

Bean, J. C. (2011), *Engaging Ideas: The Professor's Guide to Integrative Writing, Critical Thinking, and Active Learning in the Classroom*, Jossey-Bass, San Francisco.

Bok Center for Teaching and Learning at Harvard, http://isites.harvard.edu/icb/icb.do?keyword=k1985&pageid=icb.page11800.

Filene, P. (2005), *The Joy of Teaching, a practical guide for new college instructors*, University of North Carolina Press, Chapel Hill.

Science Education Resources Center, http://serc.carleton.edu/index.html.

Stanford Center for Teaching and Learning, http://ctl.stanford.edu/Tomprof/index.shtml.

Writing: Articles and Books

Clark, R. P. (2008), *Writing Tools: 50 Essential Strategies for Every Writer*, Little, Brown and Company, New York.

Silva, P. J. (2007), *How to Write a Lot: A Practical Guide to Productive Academic Writing*, APA, Washington, DC.

Strunk, W. Jr., and White, E. B. (1918, first, 1999 fourth), *The Elements of Style*, numerous publishers (Allyn and Bacon). Online and searchable version: http://www.bartleby.com/141/

Zinsser, W. (1976), *On Writing Well: The Classic Guide to Writing Nonfiction*, HarperCollins (also pdf at http://hnguyen.files.wordpress.com/2011/03/on-writing-well.pdf, and available as an audio book).

Writing: Online Coaching and Support

Academic Ladder, originally a writing club, http://www.academicladder.com/.

National Center for Faculty Development and Diversity, http://www.facultydiversity.org/.

12

MENTORING PHYSICAL OCEANOGRAPHY WOMEN TO INCREASE RETENTION

Susan Lozier and Sarah Clem

Nicholas School of the Environment, Duke University, Durham, North Carolina

ABSTRACT

The proportion of women receiving their PhD in physical oceanography has approached 35–40% at most major oceanographic institutions; however, the number of women with principal investigator status remains fairly low. A 2005 survey of 16 universities/institutions, as well as two government laboratories, found that women comprise 19% of the physical oceanographers in associate-level positions—a position assumed to be held by those who attained their PhDs between 1991 and 1999. Enrollment data from Joint Oceanographic Institutions (JOIs), averaged from 1988 to 2001, show that women constituted nearly 35% of oceanography graduate students. These statistics imply that the retention rate for women was half that for men over this time period. This chapter describes the Mentoring Physical Oceanography Women to Increase Retention (MPOWIR) program, which focuses on the role of mentoring in the early career stages of a young scientist. The primary objectives of MPOWIR are to (1) provide continuity of mentoring from a young woman's graduate career, through her postdoctoral years to the early years of her permanent job; (2) establish a collective rather than an individual responsibility within the physical oceanography community for the mentoring of junior women; (3) provide a variety of mentoring resources and mentors for a range of issues; (4) cast a wide net to avoid exclusiveness; and (5) open this program to all those who self-identify as a physical oceanographer. The impact of MPOWIR so far has been very positive toward improving retention of women.

Women in the Geosciences: Practical, Positive Practices Toward Parity, Special Publications 70. First Edition. Edited by Mary Anne Holmes, Suzanne OConnell, and Kuheli Dutt.
© 2015 American Geophysical Union. Published 2015 by John Wiley & Sons, Inc.

12.1 Background

Efforts over the past several decades toward increasing the number of women entering science and engineering have largely been successful, with undergraduate and graduate school enrollments averaging between 30% and 50% women [*Nelson*, 2002]. PhD attainments show similar progress. However, the percentage of women occupying tenure-track positions has not risen commensurably. Across the board, women in science and engineering fill, on average, only 15%–25% of academic positions [*Nelson*, 2002]. Because the number of women in graduate school has been sufficiently large for at least a decade, it is difficult to ascribe the lower percentage of women in entry-level faculty positions to a small pool of potential candidates. Thus, while recruitment efforts should be lauded, we need to also turn our attention to retention if we are to capitalize on the investment that funding agencies and universities have made on the education of women students, and, importantly, if we are to create a scientific workforce whose diversity more closely matches that of the student population and, in a broader sense, that of the U.S. population as a whole.

Ocean sciences provide no exception to these trends. The proportion of women receiving their PhD in physical oceanography has approached 35%–40% at most major oceanographic institutions; however, the number of women with principal investigator status remains fairly low. A 2005 survey of 16 universities/ institutions, as well as two government laboratories, found that women make up 19% of the physical oceanographers in associate-level positions, a position assumed to be held by those who attained their PhDs between 1991 and 1999. Enrollment data from Joint Oceanographic Institutions averaged from 1988 to 2001 shows that women constituted ~35% of oceanography graduate students. These statistics imply that the retention rate for women was half that for men over this time period, though historical data, disaggregated by discipline and gender, is needed to compute more accurate retention rates.

The *Nelson* [2002] diversity study, as well as concerns within the community, prompted members within the physical oceanographic community to examine whether mentoring efforts could aid the retention of junior women in the field. Although institutions are increasingly focusing on the role of mentoring in the early career stages of a young scientist, it is generally recognized that a discipline-based community can also foster success during a scientist's early career. Members of the community can advise a junior scientist on a host of issues, ranging from funding sources to collaborative work to research programs. With this in mind, a group of female physical oceanographers obtained funding from the Office of Naval Research (ONR) and NSF in the spring and summer of 2004 for the purpose of investigating the retention issue of women in the field of physical oceanography and how mentoring might aid such retention. This initial group originated the concept for MPOWIR (Mentoring Physical Oceanography Women to Increase Retention) and formed the first MPOWIR steering committee.

12.2. Formulation of MPOWIR

The MPOWIR steering committee recognized that the physical oceanography community could not simply adopt a program that has been developed from another discipline. A career in oceanography is unique in that it often requires sea time, there are few industry jobs, the number of geographical locations where oceanography jobs are available is limited, there are a relatively large proportion of research positions versus academic positions, and the field is relatively small compared with computing sciences, mathematics, physics, and so on. Thus, the aim was to design a program unique to, and designed by, the physical oceanography community. To achieve this goal, the steering committee sought input from a broad segment of the community by hosting a workshop. The committee wanted representation from the spectrum of workplaces for physical oceanographers, as well as from different career stages. Finally, but most importantly, the steering community decided to invite men to participate in this workshop. Men have been providing the bulk of mentoring in the field for many years, and the committee believed it was important to gain from their experience in this endeavor. Such inclusion is also a statement that the committee believes the lack of retention for women in the field is not a "women's issue" but is instead a community issue.

Many factors contribute to the lack of retention of women scientists: competition between family building and career building, competition between career goals of spouse/partner, lack of female role models, lack of adequate mentoring, and others. While some of these problems are best met with institutional changes, the latter problem in particular is one the physical oceanographic community has decided to address. Toward this end, an NSF- and ONR-funded workshop was conducted at the Airlie Center in Warrenton, Virginia, on October 9–12, 2005. Twenty-nine physical oceanographers, men as well as women, assembled for the purpose of designing a mentoring program for junior women in the field of physical oceanography in order to help remove barriers, real or perceived, in their career development.

During this first official effort, participants followed an agenda established to (1) gather input from the participants that would aid a collective identification of the particular needs for mentoring within the physical oceanographic community, (2) design a mentoring program that would meet those needs, (3) set up an implementation strategy for the program, and (4) establish metrics for measuring the success of the program. MPOWIR workshop participants identified gaps and needs for mentoring that are not currently filled by local institutional mentoring or by the peer mentoring afforded by PODS (Physical Oceanography Dissertation Symposia: http://www.pods-symposium.org/).

The initial focus at the workshop was on identifying the obstacles that junior women face in their career development and deciding which of those obstacles could be met by a community-based effort rather than by institutional efforts. A community-wide survey conducted prior to the workshop provided important input for the mentoring program design. From the survey and workshop discussions, it was concluded that transitions from PhD to postdoc and then from postdoc to entry-level positions were the most vulnerable times for a junior woman in the field. Identified obstacles included exclusion from large programs, lack of collaboration and collaborators, lack of senior women role models, and lack of advice on career development and on balancing work and family. Importantly, the survey results showed that only 30% of the respondents formed a significant mentoring relationship during their postdoctoral years.

To make mentoring accessible to junior women in a wide variety of positions and at different types of workplaces (e.g., research institutions, government labs, universities, industry), workshop participants decided on a multiprong approach with several elements, including a workshop dedicated to mentoring, mentoring groups that meet monthly, and workshops and socials at national meetings. All program elements are described in detail below. Further information on the Airlie workshop can be found in the workshop report [*Lozier et al.*, 2005] and in the meeting report published in *EOS* [*Lozier*, 2006].

Participants at the MPOWIR workshop designed a community mentoring program that would provide continuity from the PhD attainment through the early years of a young woman's scientific career. Importantly, the workshop participants decided to focus on the collective community responsibility for mentoring rather than on mentoring that matched a single junior scientist with a single senior scientist. The working hypothesis is that a network of mentors would better fulfill the various needs of a junior scientist. Following the workshop, funding from NSF, DOE (Department of Energy), NASA, and ONR was secured for the implementation of MPOWIR program activities in the spring of 2007.

12.2.1. Goals of MPOWIR

The primary objectives of MPOWIR are to (1) provide continuity of mentoring from a young woman's graduate career, through her postdoctoral years to the early years of her permanent job, (2) establish a collective rather than an individual responsibility within the physical oceanography community for the mentoring of junior women, (3) provide a variety of mentoring resources and mentors for a range of issues, (4) cast a wide net to avoid exclusiveness, and (5) open this program to all those who self-identify as a physical oceanographer. Each of these goals is intended to make mentoring opportunities universally available and of higher quality by expanding the reach of mentoring opportunities beyond individual home institutions. For the purposes of this initiative, retention is defined as continued employment within the field of physical oceanography. Since the program is focused on retaining junior women through the transition from graduate school to a postdoctoral position, and then from a postdoctoral position into a permanent appointment, retention is gauged by the percentage of women who remain in the field several years into their first permanent position.

12.2.2. Program Elements

12.2.2.1. Pattullo Conference. The centerpiece of the MPOWIR program is the Pattullo Conference, named for June Pattullo, the first woman in the United States to receive a PhD in physical oceanography from Scripps Institution of Oceanography in 1957. The first Pattullo Conference was held May 18–21, 2008, in Charleston, South Carolina. Since 2008 there have been Pattullo Conferences in 2010, 2011, and an upcoming conference is planned for 2013. The main goals of the Pattullo Conference are as follows:
- To provide junior women with career advice and feedback on their research
- To build community networks with peers and senior scientists
- To build confidence and skills for promoting one's research
- To raise awareness of issues confronting junior women among the senior scientist community

A variety of session formats are utilized to meet the conference goals. Participants give short research talks on which they receive feedback from other junior and senior scientists. This, in combination with professional development sessions (i.e., negotiations and proposal writing) strengthens participants' confidence and skillsets. Participants actively build networks through meeting in small groups and participating in one-on-one mentoring. Informal interactions at meals and during free time provide additional opportunities for organic conversation and mentoring.

Overall, the Pattullo Conference has reached 75 junior women and 41 senior scientists representing 52 institutions. Based on feedback gathered from the participants, the Pattullo Conference is a valuable experience for everyone involved and has been an extremely successful event. In follow-up surveys administered after each conference, nearly every participant said they would "definitely recommend this conference to another junior scientist" (see Table 12.1). Most importantly, the conference goals were accomplished. In evaluations and in conversation, many junior women spoke of increased confidence and were impressed by the networking opportunities with not only senior scientists but also their peers. One junior participant commented, "I am leaving with more confidence in myself and a much better idea of where I want to go in my career and why I want to do it." Another participant remarked, "This was a very helpful experience for me as a junior scientist and has definitely increased the likelihood that I will stay in the field." Many participants state the immediate, tangible benefits of the conference as well: "It provides an opportunity for you to build up your research network, to learn how to apply for funding, and how to manage your time among research and life, etc."

Table 12.1 Post-Pattullo Survey (2008, 2010, and 2011) of junior and senior scientists.

	Average Junior Scientist Response	Average Senior Scientist Response
Please rate on a scale of 1–5 (poor–excellent)		
Networking opportunities	4.78	
Professional development opportunities	4.62	
Feedback on research	3.72	
Please rate on a scale of 1–5 (strongly disagree–strongly agree)		
My skills and expertise were used to their fullest		4.22
My time was well spent at this conference		4.78
I had enough information/background about the conference to participate fully		4.54
I would attend another Pattullo Conference		4.78
Please rate on a scale of 1–5 (not valuable–extremely valuable)		
Value to current position	4.62	
Value to future position	4.6	
Overall value	4.69	
Perceived value of conference to a junior scientist		4.76
Perceived value of conference to another senior scientist		4.3
Please rate on a scale of 1–5 (definitely not–definitely)		
Would you recommend this conference to another junior scientist?	4.95	
Would you recommend this conference to another senior scientist?	4.61	4.73

12.2.2.2. Mentor groups. To keep the momentum generated by the first Pattullo Conference in 2008, mentoring groups were established the following fall. The mentoring groups are intended to support both peer and traditional mentoring on a smaller, more intimate basis. Each group meets monthly via conference call for approximately 60 minutes. The objectives are to help junior women make connections and gain community support, to offer junior scientists advice and strategies for professional success, and to help them learn from the experiences of both senior scientists and peers.

New mentor groups are formed approximately once a year. MPOWIR announces the opportunity to join a mentor group through e-mail lists and promotion on the MPOWIR Web site. New members are drawn from junior participants from Pattullo Conferences as well as women who have learned about MPOWIR through our Web presence, attendance at town hall sessions, and through colleagues and collaborators. As part of the registration process, background information on research interests, educational level, and affiliations is collected. Groups are matched to maximize similarities in career stage and to minimize overlap with individuals at their home institution. Groups consist of six to nine junior women and two senior facilitators. As of the summer of 2013, six mentoring groups were operational. Groups meet for two years, at which point members have the option to join another mentor group or form a peer group. Peer groups continue with the same format for calls but without the leadership of senior scientists. MPOWIR facilitates peer group meetings for as long as they wish to continue.

To ensure that the groups offer an immediate tangible benefit to attendees, the junior women are asked to formulate specific goals that they wish to work toward during the coming year. Prior to the first meeting of the mentoring group, each member and mentor leader received a notebook containing the biography and goals for each participant. These goals, along with other topical issues, are discussed during the mentoring group calls. Based on the 2011 survey, 100% of mentor group participants reported that they made progress on their stated scientific, professional, and personal goals. After each call, relevant articles, Web links, and information are often shared with the broader community through the MPOWIR blog.

To evaluate the effectiveness of the mentor groups, MPOWIR conducts an annual survey. Participants are asked what they value about their mentor group, the effect of being in a mentor group in their current position (Figure 12.1), and questions about the logistics and setup of the groups. Based on the survey conducted in 2012, participants all rated mentor group participation as a valuable experience, with particular emphasis on feedback about professional development and on personal matters (Figure 12.2).

12.2.2.3. Web site. The MPOWIR website (www.mpowir.org) is the central place for information on all MPOWIR activities and is intended to provide mentoring opportunities to all physical oceanographers regardless of gender or home

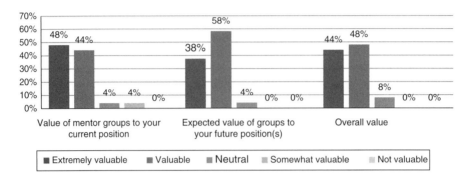

Figure 12.1 Results of MPOWIR's 2012 survey to evaluate the effectiveness of mentor groups. For color detail, please see color plate section.

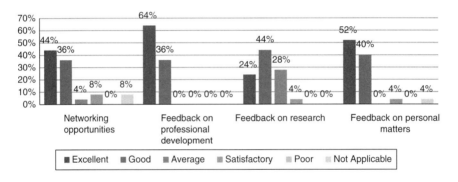

Figure 12.2 Results of MPOWIR's 2012 survey showing value of mentor groups to feedback on professional development and personal matters. For color detail, please see color plate section.

institution. As was clear from the geographical distribution of the Pattullo Conference and Mentor Group participants, many junior women are isolated in departments that are not traditionally considered to be "oceanographic." While this provides an opportunity for fruitful interdisciplinary research, it can also be isolating and create difficulties for their involvement in larger research programs. The Web resources are intended to bridge the geographical divides, to provide information and resources, and to encourage peer and traditional mentoring opportunities. Over the past two years (2010–2012) the Web site has averaged 1,025 visits per month. The resources currently provided on the MPOWIR Web site include the following:

- Profiles of female physical oceanographers, included to illustrate different career paths and provide a context for women in the physical oceanographic community
- Information, registration, and photos from the Pattullo conferences

- Resources associated with tracking, mentoring, and encouraging the participation of women in science
- A blog, which is proving to be an effective venue for the exchange of information and ideas and where job opportunities and articles relevant to women in science are shared

12.2.2.4. Town hall meetings and socials. MPOWIR town hall meetings and socials are intended to facilitate networking between physical oceanographers, to provide early career advice and career development information to junior scientists, and to enhance the sense of community. The town hall events are open to all attendees at the ocean sciences meetings, and as such, serve to broadcast the MPOWIR effort and to engage a wider audience. To date MPOWIR has sponsored five events at national meetings:

- *Town hall meeting at Ocean Sciences in 2006*: At Ocean Sciences 2006 in Honolulu, Hawaii, MPOWIR held a town hall meeting to introduce MPOWIR to the oceanographic community. Over 150 people, men and women, attended this informational meeting. At the meeting, a panel of Airlie workshop participants presented the goals and planned activities of the MPOWIR program.
- *AGU Fall Meeting reception in 2007*: At the AGU Fall Meeting in San Francisco in 2007, MPOWIR hosted a social for the purpose of providing networking opportunities for junior researchers in the field.
- *Town hall meeting at Ocean Sciences in 2008*: MPOWIR sponsored a social at the 2008 Ocean Sciences Meeting in Orlando, Florida, on dual-career couples. Dr. Elizabeth Creamer, a professor and researcher of issues related to faculty careers at Virginia Tech, spoke about her research to approximately 75 people. The background and career-stage of the audience was varied, and while many of the audience members were physical oceanographers, scientists from all of ocean science were in attendance. Dr. Creamer presented statistics on the number of dual-career couples in the sciences and summarized recent research about the impact of children and an academic spouse on faculty research productivity.
- *Town hall meeting at Ocean Sciences in 2010*: MPOWIR sponsored a town hall meeting at the 2010 Ocean Sciences Meeting in Portland, Oregon. The meeting, which was titled "Where Do All the Oceanographers Go? Career Paths in Oceanography," featured a panel discussion exploring career paths taken by oceanographers over the years. It focused on current opportunities for graduates with PhDs in oceanography. There was standing room only at this event.
- *Town hall meeting at Ocean Sciences in 2012:* MPOWIR hosted two town hall events at the Ocean Sciences Meeting in Salt Lake City, Utah, in collaboration with AWIS (Association for Women in Science). Both events

drew on the recent implementation of NSF's Career-Life Balance Initiative. The first event was a panel discussion featuring Joan Herbers (AWIS), Eric Itsweire (NSF), and Debra Bronk (Virginia Institute of Marine Science). Each focused on a different facet of balancing work and life. The second event was an informal discussion about work-life balance.

12.2.2.5. NOAA internship. Since 2009 MPOWIR has collaborated with NOAA to provide internship opportunities for graduate students. The goal of the NOAA/MPOWIR internship program is to familiarize junior women in the field of physical oceanography with the research conducted at the NOAA labs and to afford NOAA scientists the opportunity to work with a graduate student on a project of joint interest. Each year, two junior scientists are chosen for an internship at AOML, GFDL, or PMEL (Atlantic Oceanographic and Meteorological Laboratory, Geophysical Fluid Dynamics Laboratory, Pacific Marine Environmental Laboratory). Students are integrated into an ongoing program of mutual interest for a period of 8 to 10 weeks and are mentored by a NOAA researcher. Prior to the start of the internship, the students communicate with their supervisor about their goals and ideas for their project with the aim of a coauthored publication. This opportunity is open to any female scientist who is currently enrolled in a graduate program.

Participants see the internship as an opportunity to meet and work with experts in their field. Interns often comment on the lasting effects of the internship on their career, stating, "The effects of the MPOWIR internship will undoubtedly continue to influence my future work."

12.2.2.6. NASA Speaker Series. The goal of the NASA MPOWIR Speaker Series is to familiarize junior women in the field of physical oceanography with the research conducted at the NASA labs and to inform NASA scientists of the research conducted by junior scientists in the community. Each year, two scientists are chosen to give a seminar at either Jet Propulsion Laboratory or Goddard.

Since 2009 eight junior scientists have participated in the NASA Speaker Series. A past participant summarized the speaker series saying, "NASA MPOWIR Speaker Series was beneficial for allowing me the opportunity to learn about Goddard community and to present my research to them. I left my visit feeling both inspired by the conversations that I had and grateful for the gained insights for how I might better contribute to discussions in similar experiences that I may encounter in the future."

12.2.2.7. Statistics and surveys. Beginning in 2005, MPOWIR has gathered gender and discipline-specific information on enrollment and PhD attainment since 2001 from 25 degree-granting oceanography programs. The gender-specific data gives a concrete measure of graduation rates and retention in the field. As with data collected by NSF and by other researchers, our data shows that nearly 40% of all PhD graduates in physical oceanography are female, and the percentage of women in the field drops sharply as they progress in their careers.

MPOWIR also conducts a comprehensive survey of all graduate students, male and female, enrolled in physical oceanography graduate programs across the country. This survey asks students to reflect on their graduate experience and their attitudes about the field of physical oceanography. While the survey is anonymous, the majority of respondents provided contact information, which will allow us to track their progress and changing attitudes through their early career. Results of two questions from the 2008 canvas illustrate the type of information that we are gathering. As seen in Figure 12.3, men and women responded differently when asked about their advisor relationship and overall graduate student experience, with more women than men, proportionately, reporting negative opinions. While approximately the same proportion of women and men have mentors, women are more likely to have a mentor who is not their academic advisor. Whether

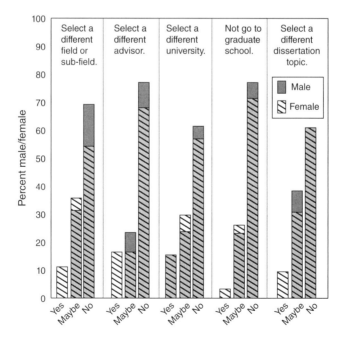

Figure 12.3 Women are more likely to report that they would change decisions related to their graduate studies.

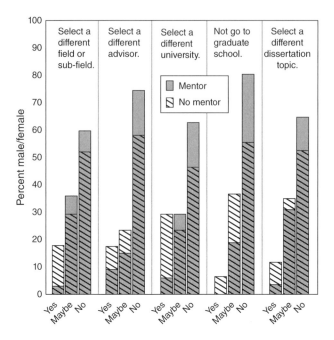

Figure 12.4 Nearly 40% of both male and female respondents do not have a mentor. Whether or not a student has a mentor significantly affects their attitude toward graduate school. For color detail, please see color plate section.

or not a student has a mentor also affects his or her response. Figure 12.4 suggests that those without mentors have a more difficult graduate school experience and are much more likely to report that they would change decisions related to graduate school.

Once we have repeated this survey for sufficient years to track any evolving attitudes about the field of physical oceanography, we plan to publish the full survey results. These results, and results from future surveys, will help guide our mentoring efforts.

12.3. Assessment of MPOWIR

To date, the impact of this proposed work has been measured qualitatively, namely through the participant surveys discussed above. However, since the overall goal of the program is to increase retention, the metric of success is clearly a quantification of retention improvement. At the start of this program, in 2007, a survey was taken of all universities across the country with oceano-graphic departments. An assessment was made of the gender breakdown at the

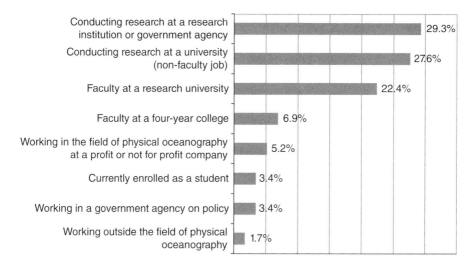

Figure 12.5 Part I of MPOWIR participant survey. Junior women were asked to indicate their current position.

assistant, associate, and senior scientist position levels. We plan to conduct a similar survey at the 7-year mark of **MPOWIR** activities, the spring of 2015 (the first Pattullo Conference and the formation of the first mentoring groups were in 2008), to assess the degree to which women have moved into the ranks of assistant scientist/professor ranks. We plan to use the field of chemical oceanography as a control, since it has not had a mentoring program during these past years.

However, this spring, at the 5-year mark of **MPOWIR** activities, a survey was sent to all junior women who had participated in *Pattullo* [2008 or 2010], and/or in mentor groups starting in 2008, 2009, and 2010, in an effort to assess the impact of **MPOWIR** on their retention in the field. All of these women joined **MPOWIR** somewhere between 3 and 5 years ago. Of the 65 participants contacted, 61 responded, resulting in an 89.7% response rate. Of the 61 respondents, only one woman is currently working outside the field of oceanography in an unrelated field (Figure 12.5). Two are currently enrolled as students. All of the others are effectively in the field, with a remarkable 79% conducting research at a university or research institution.

Since the pipeline has historically been "leakiest" at the post-PhD transition, we believe that such retention shows remarkable progress. As for what might explain this retention, Figure 12.6 shows that the participants ranked **MPOWIR** as having the greatest impact on professional development skills and professional networking opportunities.

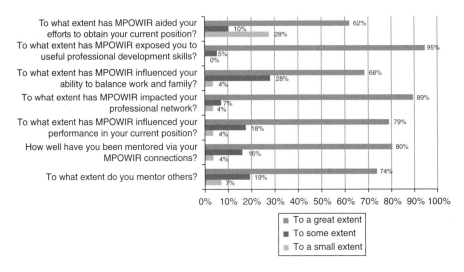

Figure 12.6 Part II of MPOWIR participant survey. Junior women were asked to indicate the overall impact of MPOWIR on their careers. For color detail, please see color plate section.

In conclusion, MPOWIR has seen many anecdotal, yet very positive, responses to programs and offerings, and a recent survey of past participants shows remarkable retention. In the near future, we plan to monitor retention by assessing the percentage of early-career women holding positions in physical oceanography at universities and institutes around the country.

REFERENCES

Lozier, M. S. (2005), A community effort toward the retention of women in physical ocean- ography, *Oceanography*, *18*, 35–38.

Lozier, M. S., A. Adcroft, L. Beal, A. Bower, D. Byrne, A. Capotondi, V. Coles, K. Lavender, C. M. Lee, J. MacKinnon, K. Shearman, L. St. Laurent, L. Thompson, and D. Witter (2006), MPOWIR: Mentoring physical oceanography to increase retention, Report of a workshop held October 9–12, 2005. Available at http://www.mpowir.org.

Nelson, D. J. (2002), A national analysis of diversity in science and engineering faculties at research universities. Retrieved from http://chem.ou.edu/~djn/diversity/briefings/ Diversity%20Report%20Final.pdf.

13

ASCENT, A DISCIPLINE-SPECIFIC MODEL TO SUPPORT THE RETENTION AND ADVANCEMENT OF WOMEN IN SCIENCE

A. Gannet Hallar[1], Linnea Avallone[2], Heather Thiry[3] and Laura M. Edwards[4]

[1] Storm Peak Laboratory, Division of Atmospheric Science, Desert Research Institute, Steamboat Springs, Colorado
[2] Division of Atmospheric and Geospace Sciences, National Science Foundation, Arlington, Virginia
[3] Ethnography & Evaluation Research, University of Colorado Boulder, Boulder, Colorado
[4] South Dakota State University Extension, Aberdeen, South Dakota

ABSTRACT

Atmospheric Science Collaborations and Enriching NeTworks (ASCENT) was a workshop series designed to bring together female scientists in the field of atmospheric science and related disciplines. ASCENT was a multifaceted approach to retaining female junior scientists through the challenges in their research and teaching career paths. During the workshop, invited successful senior women scientists discussed their career and life paths. They also led seminars on tools, resources, and methods that can help early career scientists to be successful. Networking was a significant aspect of ASCENT, and many opportunities for both formal and informal interactions among the participants (of both personal and professional nature) were blended in the schedule. The workshops were held in Steamboat Springs, Colorado, home of a high-altitude atmospheric science laboratory, Storm Peak Laboratory, which also allowed for tours with scientific engagement and a pleasant environment for participants. Near the conclusion of each workshop, junior and senior scientists were matched in mentee-mentor ratios of two junior scientists per senior scientist. An external evaluation of the workshop participants concluded that the workshops have been successful in establishing and expanding personal and research-related networks, and

Women in the Geosciences: Practical, Positive Practices Toward Parity, Special Publications 70.
First Edition. Edited by Mary Anne Holmes, Suzanne OConnell, and Kuheli Dutt.
© 2015 American Geophysical Union. Published 2015 by John Wiley & Sons, Inc.

that seminars have been useful in creating confidence and sharing resources for such things as preparing promotion and tenure packages, interviewing and negotiating job offers, and writing successful grant proposals.

13.1. Introduction

There is substantial evidence that women, as a group, are underrepresented in senior academic ranks within the science, technology, engineering, and mathematics (STEM) research fields [e.g., *Hill et al.*, 2010; *Holmes et al.*, 2008; *Everett et al.*, 1996]. Due to an increase in women graduates entering the science and engineering fields in recent years, the average age of women faculty members is lower than the men faculty members. However, even after accounting for the lower number of women than men available, many studies report that women are less likely to appear in senior academic ranks, as women faculty have lower rates of promotion [*NSF*, 2003; *Kulis et al.*, 2002]. Underrepresentation of women in faculty ranks is especially noticeable within geosciences; in 2008 females made up 14.2 percent of tenure track faculty in geosciences departments compared to 28 percent in tenure-track positions in all science and engineering fields [*AGI*, 2009]. At the same time, about 41% of the graduate students in the geosciences were female. Additionally, women are more concentrated at nonresearch institutions than research universities. For example, 18% of the faculty (both tenure track and nontenure track) at two-year institutions were women; at institutions granting bachelor's and master's degrees, 14% of the faculty were women; at PhD-granting institutions, only 10% of the faculty were women in 1996–1997 [*Macfarlane and Luzzadder-Beach*, 1998; *Ongley et al.*, 1998]. In 2008 within the geosciences, 18.6 percent of nontenure track positions were held by women [*AGI*, 2009]. The level of female participation in faculty positions has not changed significantly in recent years [*AGI*, 2009]. Also, retaining women graduate students, especially within physical sciences and engineering, has proven to be an ongoing challenge for many institutions [*NRC*, 2006]. One reason for attrition is a sense of isolation because there are so few other women in their area of study [e.g., *Whitten et al.*, 2003], and many women faculty members cited feelings of isolation as a major reason for their departure from academia. In response to this situation, the *National Research Council* [2006] has stressed the need for a policy response to create mechanisms for bringing female postdoctoral researchers, junior scientists, and senior faculty together.

Barriers confronting advancement of women scientists are complex. Factors include lack of women mentors and role models, lack of critical mass, and isolation from collegial networks [*Zuckerman et al.*, 1991]. Women role models can have significant influence on women students, particularly as they begin thinking about career choices [*NRC*, 2006]. Providing opportunities to develop professional networks is also a valuable means to combat isolation and to improve career prospects. *Bozeman and Lee's* [2005] study of 1370 random samples from university professors and researchers who are affiliated with NSF and Department of Energy centers in U.S. universities showed conclusive evidence that the number of collaborators remains the strongest predictor of productivity, as measured by publication rate. In fact, more collaboration outside of one's own work group (e.g., persons in other universities, other nations) is associated with being male, a tenured faculty member, and an increase in total number of publications. One underlying cause of challenges facing academic women is an historically male-oriented departmental culture, which can lead to isolation and discrimination, limiting the potential for research collaborations with other women scientists [*NRC*, 2006]. Academic isolation includes exclusion from access to informal sources of professional information that "are indispensable to professional development, career advancement and the scientific process" [*Etzkowitz et al.*, 1994].

In a report sponsored by NSF and authored by *Holmes and OConnell* [2004], who used the 2002 AGI Directory of "self-reporting specialties," atmospheric sciences/meteorology ranked last in the percentage of females in a tenure track position at a PhD-granting institution. This follows a previous report [*NSF*, 1997] that noted women remain least well represented in atmospheric sciences within the subfield of geosciences. From a recent 2007 report issued by The Chronicle of Higher Education, we used the Faculty Scholarly Productivity Index of top ten departments for atmospheric/meteorology, analyzed faculty rosters in those departments, and tabulated total faculty compared to the total number of women faculty. According to *Nelson* [2005], women need to be at least 15% of an organization in order to start impacting that organization's culture, policy, and agenda. Even when combining all ranks, women still compose a small fraction of the faculty of atmospheric science/meteorology departments, with three of the top 10 programs well below the 15% mark [*Avallone et al.*, 2013]. Several of these departments have just reached "critical mass," which may facilitate institutional change. Despite growth in both the number and percentage of women entering the field [*Hartten and LeMone*, 2010], women in atmospheric sciences are among the most underrepresented in tenure-track positions. The recent NRC report on research-based doctorate-granting institutions [*Ostriker et al.*, 2010] gives the mean percentage of female faculty in the top ten atmospheric science/meteorology departments as 15.1%, with a range of 0% to 50%, and a standard deviation of 8% [*Avallone et al.*, 2013].

Additionally, the largest discrepancy in salary by gender was found in atmospheric sciences/meteorology for the "university/college" employment category [*Zevin and Seitter*, 1994]. This trend was further investigated by *Winkler et al.*, [1996], and they found the most dramatic gender difference was evident at the senior level for academic employees in atmospheric sciences. These salary discrepancies have been identified for women scientists in general [*Holden,* 1993], yet the salary discrepancy for women in atmospheric sciences appeared to be significantly larger [i.e., $18,000 in atmospheric sciences compared to $7,000 in all scientific fields; *Curtin and Chu,* 1993]. *Winkler et al.* [1996] also found that women faculty within atmospheric sciences were not progressing as steadily as male faculty members in promotions. In fact, relatively more women than men were dropping out at each level. While *Hartten and LeMone* [2010] report that there has been significant growth in the number of tenure-track female faculty in the atmospheric sciences and the geosciences in general since the 1990s, comparison of the percentage of women pursuing graduate degrees with the percentage of female faculty shows that women have not yet reached parity in academic positions.

In view of the general situation facing women in sciences and specifically the remarkably low numbers of women in atmospheric science faculty positions, we developed a program with support from the NSF ADVANCE program. Based on the idea that women role models can have significant influence on younger women, particularly as they make career choices [*NRC,* 2006], we created a series of opportunities to network female atmospheric sciences faculty members at different stages in their careers, as well as postdoctoral researchers, to help address these barriers.

13.2. Description of ASCENT

Atmospheric Science Collaborations and Enriching NeTworks (ASCENT) is a program focusing on women in atmospheric science/meteorology and is designed to initiate positive professional relationships among female scientists at different stages in their careers, from postdoctoral researchers to senior scientist/full professor. ASCENT was designed to achieve the following specific goals:

1. Ensure that junior women scientists know about and have access to resources and people who can help guide them through their career and life path.
2. Encourage positive mentorship and create mentoring opportunities.
3. Learn and teach others about primary obstacles for women in atmospheric sciences and meteorological fields, and develop or share communication tools to assist in navigating these obstacles.
4. Meet potential scientific collaborators at other institutions.

The ASCENT program comprised three annual workshops in Steamboat Springs, Colorado, and reunion events at the fall American Geophysical Union (AGU) meeting and the annual American Association for Aerosol Research (AAAR) conference. Each workshop was attended by 20–22 early-career female atmospheric scientists and about 10 senior scientists drawn from academe, government laboratories, and federal funding agencies. ASCENT was oversubscribed each year, and we were unable to accept all of the very qualified applicants. The workshops consisted of both formal and informal activities to promote networking among the participants, plus breakout sessions to discuss issues related to career progression and to provide training on important career skills. By fostering relationships among women faculty and researchers, ASCENT intends to develop research opportunities for participants. Over the course of the program, these women represented 61 different universities and research facilities, from 28 states and 6 countries.

While networking with like-minded women scientists, participants had the opportunity to be involved in frank discussions to explore specific promising practices toward eliminating the leaky pipeline, defined by the attrition of women at different stages in their academic careers. By fostering relationships among women faculty and researchers, ASCENT intends to develop research opportunities and improve the quality of collaborative atmospheric research conducted at multiple universities and colleges.

Breakout sessions at ASCENT were selected based on input from participants. The following breakout sessions were offered at one of the ASCENT workshops. Participants had the opportunity to participate in two of the sessions. An invited senior scientist well regarded for success in this specific topic area led each of the breakout sessions, and the sessions were primarily discussion based.

1. *Creating a successful tenure packet.* This encompassed a discussion on how to manage one's tenure packet and navigate the process to meet an institution's research, teaching, and service criteria.
2. *Communicating.* Body language, tone of voice, and posture affect communication skills in ways we often do not realize. The goal of this session was to share how to better convey one's ideas and research, and effectively communicate in professional settings.
3. *Writing successful grants.* The goal of this session was to share tips and techniques used by successful grant writers.
4. *Student advising.* This session explored various ways for supporting student advisees while maintaining time for research and other professional commitments.
5. *Time management.* One aspect of building and managing a successful research program is learning how to manage one's time. The goal of this session was to help understand what it takes and how to approach this sometimes daunting goal.

The ASCENT workshop was initiated with a keynote speaker, selected for his or her academic experience in best practices towards gender parity in science departments. Each year, this keynote speech set the stage for a discussion pertaining to the retention of women in science. Each year, ten senior scientists were selected by the ASCENT committee due to their positive contributions to the study of atmospheric sciences. Senior scientists presented a 20-minute talk on both their research and challenges faced throughout their career. Approximately 5–10 minutes of the talk focused on research and 10–15 minutes focused on challenges faced.

Each early-career scientist attending ASCENT presented a poster pertaining to her current research topic. This poster session was held each afternoon of the workshop to allow participants to find common research themes and topics for future collaboration. Attendees were encouraged to bring a poster created for a previous scientific meeting. Each early-career woman was paired with a senior woman as a mentor; these relationships have continued over time via e-mail, telephone, meeting at conference reunion events, and a semiannual newsletter. Nearly 90% of participants have maintained contact with at least one participant since the summer workshops [*Avallone et al.,* 2013]. Since there are around 30–35 PhDs awarded to women in atmospheric science annually, the workshops have reached a substantial fraction of current early-career scientists.

Additionally, female graduate students in atmospheric science participated in the ASCENT workshops as support crew. These students were from the University of Utah, Hampton University, University of Colorado, University of Nevada–Reno, and Rutgers University. They assisted primarily with logistical issues (i.e., transportation, audio and visual needs) and note-taking during breakout sessions. Each student expressed a great appreciation for the opportunity to be involved with the program. They were able to attend meals and poster sessions with women faculty and postdoctoral researchers in their field.

Group activities were organized during the workshop to provide participants social networking opportunities and retreat from the daily routine of the workshop. Activities included visiting local waterfalls, a tour of the Desert Research Institute's Storm Peak Laboratory, and visiting local hot springs. Storm Peak Laboratory is located on the west summit of Mt. Werner in the Park Range near the town of Steamboat Springs in northwestern Colorado at an elevation of 10,520 ft. (3210 m) ASL. This atmospheric science laboratory has been used in cloud and aerosol studies for more than 25 years [e.g., *Borys and Wetzel,* 1997; *Hallar et al.,* 2011]. Additionally, group dinners were organized for the ASCENT group.

Several participants were not able to obtain funding from their institution to participate in the ASCENT workshops. Thus, a limited number of need-based travel scholarships were made available to participants. Upon acceptance to ASCENT, participants were able to apply for funding by submitting a brief essay explaining the sources of funding previously sought to attend ASCENT.

13.3. Recruitment

ASCENT was supported by several partner organizations, including the Earth Science Women's Network (ESWN), the Association for Women Geoscientists, and the American Association for Aerosol Research (AAAR). The ASCENT Web site (www.ascent.dri.edu) was updated to advertise availability of the program. This site also provided a portal for application and results of the ASCENT program. ASCENT was advertised via the ESWN list serve. ESWN is a peer-mentoring network of women, and most are in the early stages of their careers. ASCENT assisted the ESWN in hosting a social event at the December 2008 AGU meeting, which was used to recruit participants. This event was extremely successful, with approximately 300 attendees. Additionally, the AAAR staff supplied ASCENT with a list of all women attending their annual meeting in 2010 and 2011. They also assisted with organizing an event during the AAAR annual meeting in 2010 and 2011 to promote ASCENT. More than 60 women attended this event each year. ASCENT also advertised in *EOS*, the AGU weekly publication, and in the *Bulletin of the American Meteorological Society* in 2009. The University Corporation for Atmospheric Research provided a list of all academic programs in atmospheric science and names of the department chairs in 2009, 2010, and 2011. The chair of each of these departments was contacted regarding the ASCENT workshops, both via e-mail and a mailing, which included brochures. These brochures were created by a professional graphic artist using the logo shown in Figure 13.1. The Association for Women Geoscientists also advertised ASCENT on their Web site and newsletter. Interested scientists were asked to complete an online application form. This application asked the applicants to describe their research, academic career objectives, community involvement, and significant events or experiences that have influenced their career path. They were also asked to describe what they hoped to gain from participating in ASCENT and comment on how they would contribute to ASCENT. Senior scientists were invited to participate in the program via a formal letter from the project PIs.

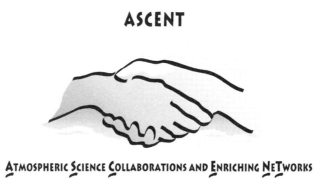

Figure 13.1 The ASCENT logo created by Lisa Wable, graphic artist at the Desert Research Institute. For color detail, please see color plate section.

13.4. Reaching a Broader Audience

To reach the general public and for recruitment purposes, ASCENT hired a video producer, Mr. Ilo Orleans from Adhoc Productions, to create short films documenting ASCENT. These films have been a success. Recent clips are available at www.ascent.dri.edu and www.youtube.com/adhocprod. During the first two weeks of streaming, this short film was viewed over 350 times. Several participants have also shown this film during faculty gathering. The film has been advertised via ESWN and the ASCENT fan page on Facebook. The ASCENT fan page was used as another mechanism to reach a broader audience and currently has 137 fans. The page is used to share updates about the workshop and recent studies pertaining to women in science.

Due to the popularity of the ASCENT film, the Desert Research Institute hired Adhoc Productions to continue this work. This internal grant funded a professional film crew and video editor to create a full-length documentary, using the video material previously collected to advertise the program. This full-length film summarizes all three years of ASCENT, and goes beyond that to reflect issues and topics for women in science. The goal of the film is bring women in science more into the spotlight for the general community, and put into place a product that will have long-lasting effects and impacts in the scientific community. The full-length film was completed in July 2012. The film features interviews with several participants from ASCENT. This film will be featured nationally at film festivals, including the Sloan Science and Film Festival and the Environmental Film Festival in Washington, D.C. The film includes seven chapters titled "Networking, Collaborations, Isolation, Storm Peak Laboratory, Obstacles, Mentors, and Life Balance."

The first ASCENT workshop resulted in two articles published in the Steamboat Pilot, a Northwestern Colorado newspaper. Additionally, a paper featuring the ASCENT model and survey results appears in the Bulletin of the American Meteorological Society [*Avallone et al.*, 2013].

The ASCENT model was also promoted via multiple presentations at the fall AGU meeting in 2010 and 2011, and via the annual ESWN newsletter in 2009, 2010, and 2011. Each year at the annual NSF ADVANCE meeting, a poster was presented on the ASCENT model.

13.5. Maintaining Contact

In addition to the Facebook fan page, ASCENT also produced an annual electronic newsletter. This newsletter focused on the successes of ASCENT participants in the previous year. This newsletter was only for ASCENT participants, and it highlighted both professional successes (e.g., publications, new position, promotions) and personal success (e.g., new child, marriage, new home) with photos and written updates. The goal of this newsletter was to help the network stay connected and promote each other.

As mentioned previously, ASCENT reunion events were held at the fall AGU meeting and the annual AAAR conference. The goal of these reunion events was to facilitate continued interaction among participants. All participants had the opportunity to meet at both the fall AGU meeting held every December in San Francisco and the annual AAAR meeting held in October. During each meeting, all ASCENT participants, including organizers, were invited to a lunch or reception to review impressions of the meeting and scientific results they found exciting. This reunion also allowed the program evaluator to log success in networking and collaboration between past participants. With this reunion, the ASCENT organizers encouraged past participants to serve as future ambassadors to recruit new ASCENT participants. Additionally, this reunion allowed for a broader networking opportunity; ASCENT members made professional contacts with both past and present participants.

13.6. Evaluation of ASCENT Success

An evaluation study was designed by Dr. Heather Thiry to provide formative feedback to program organizers about the conference design and logistics, and to gather summative information on the short- and long-term outcomes for participants. This study was conducted through the use of in-depth focus group interviews, survey instruments, participant observation at ASCENT events, and document analysis. Final findings from the post-workshop survey administered at all three ASCENT conferences are provided below, describing aggregate data from the three years. Findings on the challenges faced by women atmospheric scientists as described by focus group participants at the three summer workshops are described further in *Avallone et al.* [2013].

The quantitative survey data were entered into the statistical software package SPSS (Statistical Package for the Social Sciences), where descriptive statistics were computed. Frequencies are reported for most of the ratings items, and means for some of the items. All items were rated on a 5-point Likert-type scale (1 = strongly disagree, 5 = strongly agree). Tests of statistical significance, such as t-tests or one-way analyses of variance, were only conducted on the entire three-year data set because the small sample sizes from individual workshops precluded meaningful statistical analyses of group differences. Write-in responses to the open-ended questions and transcripts from focus group interviews were entered into NVIVO qualitative analysis software and coded as follows. Each new idea raised in a response was given a unique code name. As these same ideas were raised by later respondents, a tally was added to an existing code reflecting that idea. Frequencies of responses for open-ended items were also tabulated.

Over all three years of the ASCENT workshop, 79 women completed the postworkshop survey: 59 junior scientists and 20 senior scientists. Fifty-five of the women (70%) had never participated in a professional training similar to ASCENT.

One of the objectives of ASCENT was to provide a forum for women to discuss barriers they have encountered in their careers and to learn about the challenges faced by women in scientific disciplines. All three years, ASCENT attendees participated in focus groups that addressed the obstacles they faced; survey questions also asked about career obstacles. Work-life balance and family issues were the most frequently cited career obstacle by ASCENT participants in both focus groups and on survey questions. Women also noted isolation, not being taken seriously by colleagues, lack of institutional support, discrimination and sexism, communication issues (e.g., difficulty with negotiation, gendered communication styles, etc.), a lack of female mentors or role models, a male-oriented culture in science, and in the worst cases, intimidation and harassment. Postdoctoral researchers, in particular, faced acute obstacles. Postdoctoral researchers described a lack of access to resources and support, and the transient nature of postdoctoral positions was difficult for dual-career couples. Some women reported delaying childrearing decisions during the postdoctoral phase. Postdoctoral researchers were more likely than junior faculty members to consider leaving the field due to these challenges.

Overall, participants in all three workshops were very satisfied with the conference schedule and the variety of formal and informal activities during ASCENT. Over all three years, 95% of participants agreed or strongly agreed that they were satisfied with the design of the workshop. Additionally, 85% of all ASCENT participants agreed or strongly agreed that the mix of activities in ASCENT met their needs. In open-ended items, women reported that the specific mix of conference activities (e.g., breakout sessions, guest speaker talks, poster session, time for informal socializing, visit to Storm Peak Laboratory) helped to foster both professional collaborations and personal friendships and support networks. Given that the primary goal of ASCENT is to create networking relationships, 97% of all ASCENT participants reported that they were satisfied with the amount of time they spent interacting with colleagues at the workshop.

For the most part, the conference topics met participants' expectations and needs. Over all three years, 72% of participants agreed or strongly agreed that the breakout session topics were helpful to their professional development (the rest were neutral and no participants disagreed). Women reported that they gained valuable career tips, strategies, and advice during the breakout sessions. Participants also appreciated the open and confidential discussions in the breakout sessions.

As described above, senior scientists served as guest speakers during the ASCENT conference, discussing their research interests, personal career paths, and the challenges and successes that they had experienced as women scientists. According to survey responses, these talks were one of the most beneficial aspects of ASCENT for junior scientists, providing inspiration and motivation to overcome challenges and persist in their careers. Over all three years, 91% of junior scientists agreed or strongly agreed that the senior scientist talks were helpful to their professional development. The keynote address performed many of the same functions as the guest speaker talks in motivating and inspiring junior scientists. Additionally, the keynote address also informed participants about the status of women in science and the challenges faced by women scientists. Over all three years, 89% of junior scientists agreed or strongly agreed that the keynote address was helpful to their professional development. The vast majority of junior scientists, in survey comments and from participant observation, were appreciative of the mentoring they received from senior scientists. Many junior scientists attended the conference in order to find a female mentor in their field. Senior scientists were also significantly more likely than junior scientists to report that they gained mentoring skills from ASCENT (statistically significant, $p < .05$).

Women reported a variety of gains from the ASCENT workshop. The most frequent gain cited by both junior and senior participants was increasing their professional network; indeed, over all three years, 97% of participants reported that they enhanced their professional network. Junior scientists also gained knowledge about the issues faced by women in science and gained access to resources to help them overcome these obstacles. Ninety-three percent of all participants felt they learned about the obstacles faced by women scientists, and 90% of women felt they gained resources to overcome these obstacles. Additionally, 85% of all participants felt more confident about their future in their career after the workshop, and 87% of all participants felt more prepared to navigate their career path. Finally, 82% of all ASCENT participants reported after the workshop that they anticipated that they would collaborate with a colleague from ASCENT.

13.7. Conclusions

As an underrepresented group in atmospheric science, women face a variety of barriers to their advancement and success in the field. The ASCENT workshop provided a forum for women to discuss these issues and to develop professional and personal networks among women atmospheric scientists at varying career stages. Almost all participants reported that they enhanced their professional networks, formed personal support networks of women scientists, and gained knowledge and access to resources that will help them in their careers. Longer-term data from the workshop affirm that women maintained their networks months after the workshop [*Avallone et al.,* 2013].

While there has been significant growth in the number of tenure track female faculty in the atmospheric sciences and geosciences in general since the 1990s [*Hartten and LeMone,* 2010; *Glass,* 2011], comparison of the percentage of women pursuing graduate degrees with the fraction of female faculty shows that there is clearly still a gap and that women have not yet reached parity in academic positions. We encourage both individuals and institutions to find ways to address the factors affecting the advancement of women, as identified by women themselves, with resources available through the literature, organizations such as AWIS and AAUW, and professional societies like AGU.

As shown by *Rabinowitz and Valian* [2007], women who participate in programs similar to the one proposed here become "seeds of change in their departments" by sharing material, serving as advisors, increasing their influence, and working with administrators to improve conditions for women. Given the current representation of women within the field of atmospheric sciences/meteorology, this discipline-specific, multiple-institution program focused on retention is vital.

ACKNOWLEDGMENTS

The ASCENT program was funded via a grant from the National Science Foundation's ADVANCE program, award #HRD-0820267. The authors appreciate the support of Mr. Ian McCubbin, Ms. Jennifer Wright, and Mr. Ty Atkins for hosting the ASCENT workshops in Steamboat Springs, Colorado.

REFERENCES

American Geological Institute (AGI) (2009), Status of the Geoscience Workforce Chapter 2: Four-Year Colleges and Universities. Available at https://www.agiweb.org/workforce/reports/2009- FourYrInstitutions_rev082509.pdf.

Avallone, L., A. G. Hallar, L. Edwards, and H. Thiry (2013), Supporting the retention and advancement of women in the atmospheric sciences: What women are saying, *Bulletin of the American Meteorological Society.* doi: 10.1175/BAMS-D-12-00078.1

Borys, R.D., and M.A. Wetzel (1997), Storm Peak Laboratory: A research, teaching and service facility for the atmospheric sciences, *Bulletin of the American Meteorological Society, 78,* 2115–2123.

Bozeman, B., and S. Lee (2005), The impact of research collaboration on scientific productivity, *Social Studies of Science, 35*(5), 673–702.

Chronicle of Higher Education (2007) Faculty Scholarly Productivity Index. Available at http://chronicle.com/stats/productivity/.

Curtin, J. M., and R. Y. Chu (1993), 1992 salaries: Society membership survey, *American Institute of Physics Report,* Education and Employment Statistics Division, American Institute of Physics, One Physics Ellipse, College Park, MD.

Etzkowitz, H., C. Kemelgor, M. Neuschatz, B. Uzzi, and J. Alonzo (1994), The paradox of critical mass for women in science, *Science, 266*(October 7).

Everett, K. G., W. S. DeLoach, and S. E. Bressan (1996), Women in the ranks: Faculty trends in the ACS approved departments, *Journal of Chemical Education*, *73*(2), 139–141.

Glass, J. (2011), Increasing the recruitment and retention of women in academic geosciences: Where we are and where we should be, *AWIS Magazine*, *40*, 24–27.

Hallar, A. G., D. H. Lowenthal, G. Chirokova, C. Wiedinmyer, and R. D. Borys (2011), Persistent daily new particle formation at a mountain-top location, *Atmospheric Environment*. doi:10.1016/j.atmosenv.2011.04.044

Hartten, L. M., and M. A. LeMone (2010), The evolution and current state of the atmospheric sciences "pipeline," *Bulletin of the American Meteorological Society*, *91*, 942–956. doi:10.1175/2010BAMS2537.1

Hill, C., C. Corbett, and A. St Rose (2010), *Why So Few? Women in Science, Technology, Engineering and Math*, American Association of University Women, Washington, DC.

Holden, C., (1993), How much money is your Ph.D. worth? Careers in science, *Science*, *261*, 1810–1811.

Holmes, M.A., and S. OConnell (2004), *Where are the women geoscience professors?* Report on the workshop sponsored by NSF and AWG, Washington, DC, September 25–27, 2003.

Holmes, M. S., S. OConnell, C. Frey, and L. Ongley (2008), Gender imbalance in U.S. geoscience academia, *Nature Geoscience*, *1*, 79–82.

Kulis, S., D. Sicotte, and S. Collins (2002), More than a pipeline problem: Labor supply constraints and gender stratification across academic science disciplines, *Research in Higher Education*, *43*(6), 657–691.

Macfarlane, A., and S. Luzzadder-Beach (1998), Achieving equity between women and men in the geosciences, *Geol Soc Am Bull*, *110*, 1590–1614.

National Research Council (NRC) (2006), *To Recruit and Advance: Women Students and Faculty in Science and Engineering*, National Academies Press, Washington, DC.

National Science Foundation (NSF) (1997), *Women in Science: Celebrating Achievements, Charting Challenges*, NSF Conference Report, Washington, DC, December 13–15, 1995.

National Science Foundation (NSF) Division of Science Resources Statistics (2003), Gender differences in the careers of academic scientists and engineers: A literature review, *NSF 03-322, Project Officer*, Alan I. Rapoport, Arlington, VA.

Nelson, D. J. (2005), A National Analysis of Diversity in Science and Engineering Faculties at Research Universities, Norman, Oklahoma, January. http://cheminfo.chem.ou.edu/~djn/diversity/briefings/Diversity%20Report%20Final.pdf

Ongley, L. K., W. M. Bromley, and K. Osborne (1998), Women geoscientists in academe: 1996–1997, *GSA Today*, *8*(1), November 12–14.

Ostriker, J. P., P. W. Holland, C. V. Kuh, and J. A. Voytuk (eds.) (2010), *A Data-Based Assessment of Research-Doctorate Programs in the United States*, National Academies Press, Washington, DC.

Rabinowitz, V. C., and V. Valian (2007), Beyond mentoring: A sponsorship program to improve women's success, In A. Stewart, J. Malley, and D. LaVaque-Manty (eds.), *Transforming Science and Engineering: Advancing Academic Women* (pp. 96–115), The University of Michigan Press, Ann Arbor.

Whitten, B. L, S. R. Foster, M. L. Duncombe, P. E. Allen, P. Heron, L. McCullough, K. A. Shaw, B. A. P. Taylor, and H. M. Zorn (2003), What works? Increasing the participation by women in undergraduate physics, *Journal of Women and Minorities in Science and Engineering*, *9*, 239–258.

Winkler, J. A., D. Tucker, and A. K. Smith (1996), Salaries and advancement of women faculty in atmospheric science: Some reasons for concern, *Bulletin of the American Meteorological Society*, 77(3), 473–490.

Zevin, S. F., and K. L. Seitter (1994), Results of survey of society membership: Demographics, *Bull. Am. Meteorol. Soc.*, 75, 1855–1866.

Zuckerman, H., J. R. Cole, and J. T. Bruer (eds.) (1991), *The Outer Circle: Women in the Scientific Community*, W.W. Norton, New York.

14

FACILITATING CAREER ADVANCEMENT FOR WOMEN IN THE GEOSCIENCES THROUGH THE EARTH SCIENCE WOMEN'S NETWORK (ESWN)

Meredith G. Hastings[1], Christine Wiedinmyer[2] and Rose Kontak[3]

[1] *Department of Earth, Environmental and Planetary Sciences, Brown University, Providence, Rhode Island*
[2] *National Center for Atmospheric Research, Boulder, Colorado*
[3] *Environmental Change Initiative, Brown University, Providence, Rhode Island*

ABSTRACT

The Earth Science Women's Network (ESWN) aims to promote career development, build community, and facilitate professional collaborations for women across a variety of fields within the geosciences discipline. ESWN is a peer-mentoring network of women, with many early in their careers. ESWN started in 2002 as an initial group of six early-career women in atmospheric science and has grown to more than 1300 members. ESWN's growth has evolved solely from person-to-person contacts, and its sustained, rapid growth testifies to the group's value. The unique aspects of this disciplinary network include its focus on women at early career stages and the fact that it was formed and is led by early-career women scientists. ESWN members identify the network as a valuable part of their professional lives and often encourage peers and advisees to join. These features allow the group to assist women in the earth sciences in advancing professionally while connecting them with a community of their peers. Multiday thematic professional development workshops have been designed to bolster women's scientific career success by developing resources useful to early career geoscientists, initiating mentoring opportunities, identifying strategies for women to overcome barriers to success, and creating community. In addition, short workshops have been developed in concert with professional scientific conferences, giving access to all early-career geoscientists and enhancing useful networking opportunities. By identifying strategies to

Women in the Geosciences: Practical, Positive Practices Toward Parity, Special Publications 70.
First Edition. Edited by Mary Anne Holmes, Suzanne OConnell, and Kuheli Dutt.
© 2015 American Geophysical Union. Published 2015 by John Wiley & Sons, Inc.

reduce barriers to professional success for women geoscientists, we aim to promote a culture that will enhance the success of all scientists.

14.1. Background

In 2002, six women, all early-career scientists with similar interests and goals, started something new. They had seen each other before, at various scientific conferences and meetings; some had even collaborated on research. These women enjoyed interacting with one another, connecting on specific research topics as well as broader issues in work, career, family, and life. They counseled each other on many issues that impact the success of women in academic careers, such as balancing travel to meetings and fieldwork with family obligations and communicating effectively with graduate and postdoctoral advisors. Communicating mostly by e-mail between conferences, they discovered they had a lot to talk about.

After meeting at conferences, the six women stayed in touch and came to rely on each other for advice and connections when none were available at their home institutions. Recognizing the benefits of this informal peer network, the women slowly began including additional friends and colleagues on an e-mail list. In turn, these new invitees added their own friends and colleagues. This group of women continued to grow through "word of mouth," and the Earth Science Women's Network (ESWN) took form.

14.1.1. ESWN Today[1]

The ESWN today has become a network of women who use a variety of online and in-person pathways to connect with one another. The overall mission of the ESWN is to promote career development, build community, provide informal mentoring and support, and facilitate professional collaborations.

The ESWN currently revolves around multiple mailing lists, a Web center, social and professional networking opportunities, and career development workshops. Membership in ESWN was originally defined by joining the ESWN listserv. In 2013, a new Web center was released (www.ESWNonline.org), and this is now the primary mechanism by which women can join and interact with ESWN. Membership to the network has grown exponentially over the past 10 years (Figure 14.1), with more than 1300 members at present. ESWN is currently a women-only organization, an aspect that makes it unique and appealing to most of the members. In fact, 95% of the respondents from a survey of members in 2009 felt that ESWN should not become a co-ed group. ESWN's membership represents most major universities, government agencies, and research organizations in the U.S. and abroad. In fact, the new Web center has members logging in from 51 different countries. Membership has grown exclusively through word of

---

[1]This chapter was written in 2013; as of 2015, ESWN has over 2400 members and a newly organized Leadership Board. Please see www.ESWNonline.org for current information.

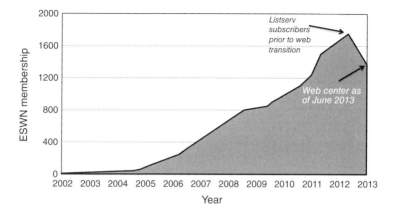

Figure 14.1 ESWN membership growth since its inception in 2002 (in 2012 the number of members is based on subscribers to the National Center for Atmospheric Research–supported listserv; June 2013 represents the number of members registered to the new Web center).

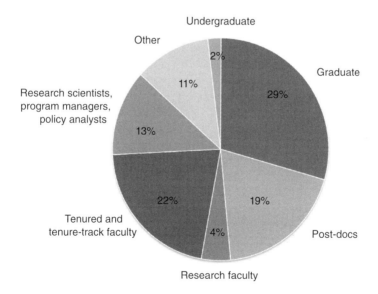

Figure 14.2 ESWN membership breakdown by (self-identified) career type.

mouth, and includes upper level undergraduates, graduate students, postdoctoral scholars (postdocs), professionals in a range of environmental fields, scientists working in federal and state governments, and professors at colleges and universities (Figure 14.2).

Until 2009, ESWN activities had been supported primarily by the volunteered time and efforts of ESWN members and a Leadership Board. As ESWN began to grow considerably, its founding members designed the Leadership Board to add structure and transparency to ESWN and its activities. Building on ESWN's history as a grassroots organization, the leadership structure is relatively "flat," with all members (eight women in 2012) of the Leadership Board contributing equally to governance of the group. Each Leadership Board member leads an area of design for the wider ESWN membership and often initiates formation of a subcommittee, drawing on the interests and concerns of the membership at large. For example, Leadership Board member Christine Wiedinmyer organized a welcoming committee in 2009, which consisted of ESWN members who directly e-mail a new subscriber to the ESWN listserv. Via a personal e-mail from a welcome committee member, new ESWN members are informed of the resources currently supported by ESWN and asked to send a personal biography introducing themselves to the whole group. Former Leadership Board member Agatha De Boer organized a subcommittee designed to organize and raise funds for activities and initiatives led by members living in the European Union.

The Leadership Board is currently organized as follows: Meredith Hastings, an assistant professor at Brown University, leads initiatives associated with the growth of ESWN and is the lead PI on an NSF ADVANCE grant (see below); Tracey Holloway, an associate professor at University of Wisconsin–Madison, oversees management of ESWN; Christine Wiedinmyer, a Scientist III at the National Center for Atmospheric Research, leads online networking activities; Erika Marin-Spiotta, an assistant professor at University of Wisconsin–Madison, oversees in-person networking, with a focus on the U.S. and the Americas; Allison Steiner, an associate professor at University of Michigan, oversees development of publications and public relations materials; Mirjam Glessmer, a postdoctoral associate at the University of Bergen, has organized professional development workshop events in Europe and is part of a subcommittee to grow membership throughout Europe; Manda Adams, an assistant professor at University of North Carolina–Charlotte, leads the development of resources for board members and reportage on Web center activity; Emily Fischer, an assistant professor at Colorado State University, leads the organization of professional development resources for ESWN's Web center; Carmen Rodriguez, a graduate student at the University of Miami, organizes the annual ESWN newsletter; and with the NSF ADVANCE support, Rose Kontak (Brown University) supports the Leadership Board activities and the ESWN mission as the ESWN Coordinator and Manager.

ESWN activities since 2002 have benefited from funding and in-kind support from the National Oceanic and Atmospheric Administration (NOAA), the National Center for Atmospheric Research, the Center for Sustainability and the Global Environment and the Center for Climatic Research of the Nelson Institute for Environmental Studies at the University of Wisconsin–Madison, the Joint Institute for the Study of the Atmosphere and Ocean in Seattle, the National Wildlife Federation, the Division of Atmospheric Science at the Desert Research Institute, the journal *Environmental Research Letters*, Brown University's

Environmental Change Initiative, University of Bergen in Norway, and the University of East Anglia in the United Kingdom.

In 2009, the ESWN Leadership Board members received an NSF ADVANCE PAID award to formally support online and in-person activities, and professional development workshops. The overall desired outcomes of this funding include increased confidence in early-career women working in the earth sciences (removing isolation, improving confidence, providing support and community); increased scientific collaborations (both nationally and abroad) for women in the earth sciences; increased retention and promotion of women in geosciences positions, with particular emphasis on advancing early career women to mid- and senior levels; and increased success of women in earth sciences through promotions, awards, publications, citations, and funded proposals.

14.1.2. Joining ESWN

ESWN is unique among organizations for women in science and engineering. ESWN has no requirements for joining other than being a woman in an earth science-related discipline: there is no membership fee or application process. This allows ESWN to effectively engage and provide community for early-career women at the graduate student and postdoctoral level and those working in the developing world, where even the smallest dues or monetary contribution can dissuade enrollment. ESWN membership is international and multidisciplinary within the earth sciences umbrella. ESWN targets women in the earth sciences who often do not fall within the membership of existing associations (e.g., in atmospheric sciences). A network that focuses on the earth sciences fosters the development of professional networks by enabling members to connect on specific research topics and develop scientific collaborations with their peers. When asked what makes ESWN different from other known groups as part of a larger survey in 2009, responding ESWN members noted the following as positive aspects of the organization:

- ESWN is personal, candid, and more direct than other groups and organizations.
- ESWN is interactive among its members via regular communication through the listserv.
- ESWN focuses on both professional and personal issues facing women.
- ESWN is international, which provides perspectives from many different cultures.
- ESWN focuses on the career development of women in earth sciences, particularly those in the early stages of their careers.
- ESWN promotes connections between its members.
- ESWN is both formal and informal.
- ESWN is free.
- ESWN is female only.

Membership in ESWN is defined by registering as a member of the Web center (www.ESWNonline.org).

14.2. ESWN Connections: Online and In Person

ESWN is a grass-roots network that promotes collaborations and interactions between women in the earth sciences. This is done through a Web center launched in 2013, a jobs e-mail list, the coordination and hosting of informal and formal in-person networking events, and professional development workshops. ESWN offers a unique space online and at in-person networking events where women earth scientists are empowered and inspired by colleagues and friends.

Prior to February 2013, an ESWN e-mail list was the primary mechanism by which members connected, asked advice, and organized in-person meetings. As of August 2012, more than 1600 email addresses were subscribed to the main ESWN mailing list, and as many as several hundred e-mails were shared among members each month. Conversation topics on the e-mail list ranged from meeting announcements, job interview tips, identifying gender issues in the workplace, and writing effective proposals to more personal inquiries, such as opinions about changing names after marriage, confronting a particularly difficult gender bias situation, caring for a child during fieldwork, and dealing with work-family balance. Members value ESWN as a women-only organization because of the nature of conversation topics initiated on the listserv, and the comfort, openness, and candor of the participants.

In addition to the primary ESWN list serve, ESWN also sponsors a secondary list where job advertisements are posted. As of 2012, the list has over 3000 subscribers, including both men and women, and approximately 70 job announcements are posted to the list each month.

A new online ESWN platform was launched in February 2013 (www.ESWNonline.org) with support from the NSF and American Geophysical Union (AGU). This resource enables ESWN to connect with more individuals and create a stronger network of dedicated women pursuing research in the earth sciences. The site includes an extensive public interface for spotlighting the accomplishments of female geoscientists, advertising upcoming events and workshops, highlighting ESWN-related news, and sharing gained knowledge from, for example, professional development workshops and collected recommendations from members. The Web center also features career resources relevant to all early-career Earth scientists. In addition to the public site, the Web center features a members-only space for networking and collaboration. This login-required, member-driven community space allows members to view searchable profiles of ESWN members and create, join, and post comments and/or documents to discussion groups on a broad range of topics. After four months of use of the Web center, members have created 38 different discussion groups and generated 1138 unique posts related to 311 different topics.

In addition to connecting women online, ESWN has also promoted face-to-face, in-person networking events. ESWN members organize events at a variety of scientific conferences throughout the year. The Leadership Board has been able to raise funds over the last eight years to support an official event annually at the fall meeting of AGU in San Francisco, and has, in recent years, expanded to other professional meetings like the European Geophysical Union and the Geological Society of America (GSA). Official gatherings at professional meetings enable women scientists to meet one another in person. Early-career members often comment that this is especially valuable, enabling them to more easily network with other conference attendees and feel more comfortable and confident connecting with colleagues during the meeting. Members of the ESWN also coordinate in-person meetings in specific geographic regions, connecting women located closely to one another.

14.2.1. Professional Development Workshops

As part of the recent funding, the ESWN Leadership Board has created a series of workshops to address professional development needs for early-career earth scientists. Besides lacking a critical mass, many other barriers prevent women from succeeding in science and engineering academic careers [e.g., *NRC*, 2006; 2007; *Holmes and OConnell*, 2004; *Wylie et al.*, 2007]. Women are subjected to unintended and unrecognized biases that can have an ultimate impact on their productivity, advancement, and success [*Wylie et al.*, 2007]. For example, *Wennerås and Wold* [1997] showed that women applicants needed higher "impact points" than their male counterparts to achieve similar ratings for a prestigious postdoctoral award. *Wylie et al.* [2007] conclude that the removal of this type of discrimination will be extremely difficult to achieve. Ways to address this type of discrimination are to (1) recognize its existence and (2) promote both "hard" and "soft" skill development [e.g., *Gura*, 2012] to help women to promote their research and themselves successfully. As of 2013, the Leadership Board has organized four multiday thematic workshops on topics such as leadership and management skills, defining research identity, and networking and communication skills. The Leadership Board also created a series of short workshops that are provided during major geosciences conferences like the AGU fall meetings and the GSA annual meetings, with the overall goals of putting into practice skills developed via ESWN events and workshops, as well as reaching a broader geosciences audience.

14.2.1.1. Multiday thematic workshops. In December 2008, the NOAA Office of Atmospheric Research supported the ESWN Leadership Board in developing and presenting its first two-day workshop for early-career women in the earth sciences. The workshop, titled "Building Leadership Skills for Success in Scientific Organizations," included 41 participants from academia (public and private institutions). Two facilitators led trainings on successful leadership and management strategies and emotional intelligence, and the workshop also included a panel of senior professionals in various positions who spoke about their career paths, successes, and struggles, and offered advice.

Since 2008, the Leadership Board has hosted three other multiday workshops with funding from the NSF ADVANCE program. This series of workshops was designed to take place in various locations around the U.S., and NSF ADVANCE funding was used to help defray some of the travel costs for participants, both in an effort to maximize participation across the ESWN membership. The first of these workshops, held in June 2011, focused on the theme "Defining Your Research Identity." This workshop was held at the National Center for Atmospheric Research in Boulder, Colorado. The second workshop was held in June 2012 at the University of Wisconsin–Madison and emphasized the subject "Skills for Networking and Communication," while the third was held in June 2013 at Brown University in Providence, Rhode Island, on the topic "Building Leadership and Management Skills for Success." The June 2011 workshop included 50 female participants; the 2012 workshop included 70; and the 2013 workshop included 75. Following the successful model from the first workshop in 2008, these workshops included professional facilitators to address the particular themes and skill building, panel discussions with open question and answer time with the audience, in-person networking opportunities with other workshop attendees as well as local scientists, and time for sharing of research.

Pre- and postworkshop surveys were given to all participants. Overall, the response about the workshops was extremely positive and the participants of the workshops were satisfied with the quality of the workshops. Results of the post-workshop surveys in 2011, 2012, and 2013, all revealed that the workshop participants made strong gains in several aspects of professional development specifically targeted by the workshop. For example, attendees reported an increase in their clarity about their goals, preparedness to navigate their career path, clarity about their values as scientists, and ability to communicate those values; these changes were statistically significantly different between pre- and postworkshop means. One participant of the 2011 workshop in Boulder commented: "At this stage in my career it was not only useful to help develop strategies for success, but very encouraging to listen to the experience of the senior scientists and have had the opportunity to meet many of the outstanding women associated with ESWN. Thank you again, this workshop has really helped me refine my agenda and inspired me in many ways."

Participants in 2012 reported that the workshop strengths included the diversity of participants and the positive and empowering atmosphere. Attendees also reported that their networking would become more deliberate as a result of the workshop, and that membership in ESWN can be used as a means for expanding their professional network. (More results and details from the surveys can be found at www.colorado.edu/eer/research/womenpartner.html.)

14.2.1.2. Single-day, short workshops. ESWN sponsors and leads single-day "mini-workshops" that take place over a few hours and are held in conjunction with major national scientific conferences (e.g., AGU, GSA). The timing, location,

and topics chosen for these workshops are designed to maximize access to early-career geoscientists without generating additional cost associated with travel or attendance. These mini-workshops have served several purposes: they have provided ESWN leadership a testing ground for specific topics before developing them into full, 2.5-day workshops; they allow summer workshop attendees to put into practice skills they have developed as part of those events; and they allow for men to also participate in the activities. As of 2013, ESWN has organized and led six mini-workshops at various professional meetings. These activities have been primarily designed to target early-career faculty and typically draw audiences of ~40–70 participants, with 50%–60% women in attendance and 55% of all identifying themselves as non-White (i.e., Black, Hispanic, Asian, or Pacific Islander, etc.). Further, 89% of participants have ranked the overall quality of the workshops in comparison to other professional development workshops they have attended as either good or excellent.

Since 2009, ESWN Leadership Board members have organized and facilitated mini-workshops during the annual AGU fall meeting. In 2009, a workshop titled "Writing NSF Proposals and Working the NSF System" was organized with NSF program officers. This workshop has now solidified into an annual event during the AGU fall meeting ("Navigating the NSF System"), and is also cosponsored by AGU's Education and Public Outreach section. The workshop is advertised as part of the scientific program; in 2011 a similar workshop was organized during the annual GSA meeting as well. The "Navigating the NSF System" workshop has three components beginning with (1) an introduction to NSF and presentation of aspects of NSF's programs led by an NSF program officer, (2) a panel discussion with four or five NSF program officers from the Geosciences Directorate to discuss elements of successful proposals and allow for question and answer time with the audience, and (3) an opportunity for attendees to meet in small groups with the program officers in attendance (~15–20 from across the Geosciences Directorate). These three segments over the course of two to three hours are divided such that attendees may attend only portions or the entire program.

Annually, since 2010, additional topical workshops have been added. These feature expert, volunteer panelists and are dedicated to a professional development skill such as publishing tips in the geosciences, networking, time management, leadership skills, or independent research design. These topical workshops feature a four- or five-member panel of scientists who present aspects of their expertise on the subject followed by questions and answers with the audience and then networking time with participants. The agenda for each workshop is designed to allow for participants to come and go, minimizing impact on their ability to attend the scientific program of the conference.

The overall aim of hosting these workshops is to promote an expansion of the participants' professional network, awareness of models for success in scientific careers, and an increase in participants' overall confidence at seeking to stay within the sciences after completing their PhD. Based on postworkshop surveys, this "soft" skill building benefits the participants greatly, and these events tend to draw significantly from typically underrepresented populations in the geosciences.

14.3. Concluding Thoughts

ESWN is a community of scientists dedicated to supporting collaborations and providing mentorship for its members, many of whom are in the early stages of their careers. This has been a grassroots activity, driven by the needs of the members. The ESWN provides a mechanism for women to connect with one another, reducing feelings of isolation and providing resources for promoting women in their careers.

ESWN is open to any woman interested, with no application fees or process. The advantage and benefits of the network are driven by the participation and input of its members. Members are encouraged to invite to ESWN others not in the network, and joining the network is accomplished via www.ESWNonline.org. ESWN should be viewed as a resource of support for women scientists in general.

ESWN professional development workshops, both the short and the multiday experiences, have highlighted the importance of career strengthening for women at all stages of their career. Based on a survey of the membership in 2009, different types of professional growth are important at different career stages [*Laursen and Kogan,* 2010]. While expanding one's professional network was the need selected most highly overall, various groups within the network identified different needs as most important for advancing their careers: graduate students identified independent research design and building knowledge within their fields; postdocs identified independent research skills; and faculty selected time management and group management skills as most important. Overall, the participation and enthusiastic responses to the ESWN professional development workshops highlight that strengthening softer skills (e.g., communicating, networking, negotiating, management) at early career stages leads to greater potential for a successful and more satisfying career.

Comparison of survey results from ESWN members and members of the co-ed Earth Science Jobs list run by ESWN provide important additional insight into salient issues that institutions can and should address to successfully recruit, retain, and promote women scientists [*Kogan and Laursen,* 2011]. Of the ESWN members who responded to the survey, 22% marked the lack of mentors and role models as an important barrier to the retention of women in science; members rate the atmosphere in their unit less positively than do nonmember men; members rate their interactions with colleagues less positively than do nonmember men, indicating unconscious bias; 31% of members agreed that women are adequately represented in senior roles while 72% of nonmember men agreed with this, again illustrating a lack of role models; members have less-accommodating family arrangements than do nonmember men; and members spend more time than do nonmember men on their household responsibilities, complicating work-life balance. (Results from evaluation of the NSF grant activities can also be found at www.colorado.edu/eer/research/womenpartner.html.)

Additional anecdotal observations from the ESWN suggest that many women, particularly those in the earlier stage of their careers, greatly appreciate opportunities to meet and connect with other women, focus time on the promotion and development of their careers, and need support in finding mentors at all career levels. Providing monetary support for graduate students, postdoctoral researchers, and pretenure faculty to attend conferences and professional development opportunities, such as those sponsored by the ESWN, is extremely valuable and worthwhile.

ACKNOWLEDGMENTS

This publication was supported in part by funding from the National Science Foundation (ADVANCE PAID award #0929782). We gratefully acknowledge the contributions of both present and past ESWN Leadership Board members and the amazing women that are ESWN.

REFERENCES

Etzkowitz, H., C. Kemelgor, B. Uzzi, and M. Neushatz (2000), *Athena Unbound: The Advancement of Women in Science and Technology*, Cambridge University Press, Cambridge, UK.

Gura, T. (2012), Workshops that work, *Nature, 488*, 419–420.

Holmes, M. A., and S. OConnell (2004), *Where Are the Women Geoscience Professors?* Report on the workshop, September 25–27, 2003, Washington, DC.

Kogan, M., and S. Laursen (2011), Obstacles in the Advancement of Early-Career Female Geoscientists: Research Results from the Earth Science Women's Network (ESWN), ED23B-0617, presented at 2011 Fall Meeting, AGU, December 5–9, San Francisco.

Laursen, S., and M. Kogan (2010) Evaluating Career Development Resources: Lessons from the Earth Science Women's Network (ESWN), ED13A-0600, presented at 2010 Fall Meeting, AGU, December 13–17, San Francisco.

National Research Council (NRC) (2007), *Beyond Bias and Barriers: Fulfilling the Potential of Women in Academic Science and Engineering*, National Academy Press, Washington, DC.

National Research Council (NRC) (2006), *To Recruit and Advance Women Students and Faculty in Science and Engineering*, National Academy Press, Washington, DC.

Wennerås, C., and A. Wold (1997), Nepotism and sexism in peer review, *Nature, 387*, 341–343.

Wylie, A., J. R. Jakobsen, and G. Fosado (2007), *Women, Work, and the Academy: Strategies for Responding to "Post-Civil Rights Era" Gender Discrimination*, New Feminist Solutions, Barnard Center for Research on Women.

15

LEARNING TO DEVELOP A WRITING PRACTICE

Suzanne OConnell

Department of Earth and Environmental Sciences, Wesleyan University, Middletown, Connecticut

ABSTRACT

Publications are currency in the academy. The number and quality of papers are closely linked with perceived competence. Despite its importance, graduate students, postdocs, and young faculty are not usually taught effective writing practices. Indeed, surprisingly little research has been done on what promotes effective writing. How does someone become a productive writer? Why, across disciplines and countries, do women scientists have a lower publication record than their comparable male colleagues? These are some of the questions that this chapter addresses. This chapter also offers suggestions and advice on developing a successful writing practice.

15.1. The Situation

Publications are currency in the academy. The number and quality of papers is closely linked with perceived competence [*Sonnert and Holton,* 1996]. A single important paper can make a career [*Wilson,* 2012], and no publications means no academic career at the level of a four-year college and above.

How does someone become a productive writer? Why, across disciplines and countries, do women scientists have a lower publication record than their comparable male colleagues? *Fox* [1983] reviewed three categories of explanations for productivity, individual/psychological characteristics, institutional affiliations, and the reinforcing effects of feedback and cumulative advantage. No one category could explain the variation between individuals' publication

Women in the Geosciences: Practical, Positive Practices Toward Parity, Special Publications 70. First Edition. Edited by Mary Anne Holmes, Suzanne OConnell, and Kuheli Dutt. © 2015 American Geophysical Union. Published 2015 by John Wiley & Sons, Inc.

rate. Institutional prestige, however, had one of the highest correlations with publication productivity. Unfortunately, Fox offered no data about what institutional steps fostered or retarded productivity, or how or why these departments were able to identify those individuals who would become productive. Her study supported previous work [e.g., *Zuckerman and Cole, 1975*] showing the reinforcing effects of accumulated advantage in creating a scientific elite. In light of the *Massachusetts Institute of Technology women's study* [1999] showing women faculty's cumulative disadvantage, and the lower proportion of women science faculty at major PhD-granting institutions, this is a sobering situation.

More recent studies across disciplines show clear gender differences, even in fields such as the life sciences, which have the highest percentage of female faculty in natural science disciplines [*Symonds et al.,* 2006; *Eigenfactor.org*]. Women academics in linguistics and sociology demonstrated a clear productivity decline after the birth of a child [*Hunter and Leahey,* 2010]. Between 2002 and 2006, *Monroe et al.* [2008] interviewed 80 women at the University of California–Irvine. Their study identified the many subtle ways that gender discrimination is present in a professional environment that follows a very linear career path. High on their list was gender devaluation, whereby the status and power of an authoritative position are downplayed when a woman holds the position.

The pervasiveness of gender devaluation received widespread attention when *PNAS* published *Moss-Racusin et al.'s* [2012] gender study. They provided clear evidence that both men and women devalued an application for an introductory laboratory technician job when the application carried a female name. Both men and women listed the female as less able and offered her a lower starting salary. The article doesn't elaborate on why women gave both male and female applicants lower starting salaries. Could this be part of a self-devaluation?

Holleran et al. [2011] published another news-catching gender study of computer scientists and engineers. They found that men seemed energized when talking about their work. Women seemed energized when discussing their work with other women, but not when talking to men. This occurred despite the absence of any overtly hostile verbal manifestations from the men. The interpretation offered by the authors was that these discrepancies were a result of stereotype threat, the tendency of the mental baggage of negative stereotypes about a particular group to cause members of that group to behave as the stereotype predicts (see chapter 10). An NPR story [*Vedantam,* 2012] included an interview with Shirley Malcolm, Head of Education and Human Resources for the American Association for the Advancement of Science. Dr. Malcolm described the marginalization of women in scientific disciplines as a chicken and egg problem: fewer women in a field make it less likely that other women will enter the field. Negative stereotype threat for minorities (gender and ethnic) is pervasive throughout society, including in the academy, and impacts minority performance [e.g., *Steel,* 1992; *Cohen et al.,* 2006; *Walton and Cohen,* 2007, 2011].

The extent to which the depressed productivity of female faculty is linked to accumulated disadvantage or stereotype threat isn't clear. Whatever the cause, however, two positive reports suggest that female publication rates are rising along with the number of female faculty. *Wilson* [2012] pointed out, "Although the percentage of female authors is still less than women's overall representation within the full-time faculty ranks, the researchers (analyzing JSTOR data) found that the proportion has increased as more women have entered the professoriate." The correlation of increased productivity with an increase in women faculty was specifically noted in a study across all fields in Australian universities. There, between 1991 and 1993, women published 57% of the male average. Twelve years later that number had risen to 76% [*Bentley,* 2011]. This coincided with a doubling of the percentage of women in these same universities, from 21.6% of faculty in 1985 to 43.6% in 2010.

This is good news, as the number of women academics in the geosciences is also rising. Examination of the status of minority faculty is not as positive. There are few underrepresented minority PhDs in science. With a double negative stereotype [e.g., *Gutierrez y Muhs et al.,* 2012], it may take as long or longer for the academy to assimilate this poorly represented faculty group.

It doesn't have to be that way, however. Through the NSF ADVANCE program we have learned about steps to improve departmental climate, removing some of the disadvantages of "one of a kind" faculty and how to recruit a diverse faculty (see chapter 10). Hopefully, this knowledge will help to create an easier transition for future minority faculty.

15.2. Why People Don't Write

Despite its importance, graduate students, postdocs, and young faculty are not usually taught effective writing practice. Indeed, surprisingly little research has been done on what promotes effective writing. And since most faculty are not productive writers, they cannot train their graduate students in the practice of writing.

Writing is not an easy practice. Even influential, prolific writers describe how difficult it is. Boice, who has probably done the most research on productive writing, includes in his 1990 book a quote from *Virginia Valian* [1977]:

> I worked steadily, though with difficulty and anxiety; I knew, however, that I could last out five minutes of difficulty and anxiety, so I continued. At last the bell went off and I collapsed. I went to the bedroom and threw myself on the bed, breathing hard and feeling my heart race.

The reasons for not writing are extensive. As *Brande* [1934] states, "First there is the difficulty of writing at all." Web pages and bookshelves confidently display information to help a "writer-wanna-be" to become a writer, and although this advice is mostly subjective in nature, it does suggest how common the problem of writer's block is. The causes of this impasse vary, of course, but *Boice* [1990] has identified four diagnostic clusters from his extensive analysis of writing problems.

Work apprehension and low energy about writing
Depression and evaluation anxiety
Perfectionism
Procrastination and impatience

Numerous suggestions are available to help a person whose writing productivity is stuck. These include steps such as speaking the words instead of writing them, making a commitment to write at a scheduled time, imagining that you are describing the topic to a close friend or relative, and so on. Sadly, however, Boice has shown that identification of writing problems does not, by itself, lead to remediation. However, he has identified a series of steps, which if taken over several months, lead to changes in writing attitude and habits. Changing behavior is not easy, and if you consider "not-writing" a bad habit or even an addiction, this will prepare the new faculty member with some idea of what he or she is up against.

To change behavior, *Boice* [1990] has developed a program that is both rigorous and sequential. Someone who is not a productive writer may be able to arrive at an effective writing practice without following his steps, but knowing his process might help anyone to become a more productive writer.

15.3. Remedies

15.3.1. Proven Methods

Boice's [1990] guide contains questionnaires, resources, and a detailed step-by-step program to become a more productive writer. His 2000 book describes how his approach can make someone a more effective faculty member. The most convincing part of his argument is presented in a 1984 paper, in which he began to understand and develop the methods of his writing program.

Boice himself didn't have problems with writing and found that other faculty often came to him to discuss their writing problems. Later he surveyed the 10%–20% of "self-starters" who also seemed to have no problem writing. He then conducted a 10-week experiment with 27 faculty volunteers. All participants had complained about an inability to finish writing projects. At the beginning of the experiment they were instructed to schedule workday writing sessions. Then participants were divided into three groups. Group one, "Abstinence," was told to put

Table 15.1 Writing productivity and creative ideas produced by nine faculty in each control group over a 10-week period [*Boice,* 1984].

Group	No. of Pages/Day	Modal Days Between Creative Ideas
Forced Writers	3.2	1
Spontaneous Writers	0.9	2
Abstinent Writers	0.2	5

off all but emergency writing, but to keep track of their ideas. Group two, "Spontaneous," was encouraged to write only if they were in the mood. Members of the third group, "Forced Writers," were "forced" to write during all of the writing sessions. The force was applied through prewritten checks that would be sent to a hated organization on days they failed to produce three pages of writing. The results (Table 15.1) showed that, of course, the Forced Writers produced more, but that they also had more creative ideas. Their recorded responses were also positive, describing increased ease of writing as they participated in the disciplined program.

Many academics and online writing assistance programs have adapted Boice's approach. *Silvia* [2007] captures many of Boice's ideas in his succinct book, *How to Write a Lot*. The key to both Boice's and Silvia's approach to writing progress lies in (1) mental attitude (including rewards), (2) regularity of both stopping and starting, so that writers don't continue to exhaustion, and (3) accountability to someone else or another group. Through this practice and trying different techniques with faculty volunteers, *Boice* [1994] developed a six-step program to create and sustain new writing habits:

1. Motivation (Prepare in patient, timely fashion beforehand)
2. Imagination (Practice reinforcing acts of discovery)
3. Fluency (Practice writing regularly, plan fully)
4. Control (Work with mild happiness and supportive surroundings)
5. Audience (Listen to understand, then to be understood)
6. Resilience (Reinvent yourself via self-study)

These are certainly not the only way to create writing productivity, but they are a method that has worked for many academic writers.

Roy Peter Clark, a writer and teacher, has authored several writing help books that provide suggestions and solutions to writing problems. His books, *Writing Tools* [2008], *The Glamour or Grammar* [2010], and *Help! For Writers* [2011] are geared towards a more general audience, but his suggestions of creating a plan, regular practice, and curtailing self-criticism are similar to those of Boice. Silvia, author of the popular *How to Write a Lot*, offers the most compact of the books mentioned here and has similar suggestions. Find the approach that helps you the most and get writing.

15.3.2. Writing Retreats

Breaking years of bad habits that are probably equivalent to addiction is not easy. To find the mental and emotional strength to try a new approach, like the start of any new habit, may require a break from your regular routine and location, a rehabilitation of sorts. One place to get this is at a writing retreat. We briefly describe three different types.

15.3.2.1. Off campus. Beginning in 2007, weeklong summer writing retreats have been held in New England for female geoscientists (GAIN, Geoscience Academics in the Northeast). The original objective of the retreats, which were financed through the NSF ADVANCE program, was to create a supportive community for women to help them progress in their careers. At the beginning, the writing retreats were viewed as a means of community building, networking, and professional development through writing. Now, years later, learning to write and writing is the core activity. Community building and networking grow out of the writing retreat. Two videos, one long (http://www.youtube.com/watch?v=5xuTJCIDrP4) and one short (http://www.youtube.com/watch?v=P4a96nmpmWw) provide a sense of the writing retreats.

A typical GAIN writing retreat began on Sunday evening, with participants arriving after dinner. An icebreaker was held in the evening, allowing participants to meet each other and learn about the week's schedule. Food was served communally and meal times were set. Each participant had her own room and there was ample workspace throughout the facilities for writers to find their own and varied locations to write. Midmorning on Tuesday, a professional writing coach met with interested participants (there were no requirements to partake of any of the meals or activities.) These sessions varied but included discussions of writing habits and practices as well as examples of good and bad scientific writing from published papers. After lunch and before dinner, participants met individually or in small groups with the writing coach to discuss specific papers. Wednesday was devoted to writing. A program was held Wednesday evening on topics such as Getting Along with Difficult People, Social Media, and the Status of Women Scientists. Thursday was again devoted to writing, and Thursday evening there were formal and informal discussions about the week's successes and writing plans for the future.

The community-building opportunities provided by the writing retreat were fostered by invitations among participants to speak at each other's institutions, informal gatherings at professional meetings, and new research collaborations. Valuable career information was given to early-career participants by more senior participants, and those at the early stages of their careers in turn provided media/technological information to some of the more senior members of the group. In the course of 106 participant weeks (some participants have come only once, others for all six years) and 54 participants, we know of 52 papers,

2 proposals, multiple dissertation chapters, and many AGU abstracts written all or in part during the writing retreat week. Several new collaborations developed, and 17 of the participants have completed dissertations, found new jobs, or been promoted.

15.3.2.2. On campus: Weeklong. Not everyone has the luxury of being able to leave home for a week or would feel comfortable doing that. The University of Nebraska–Lincoln and ADVANCE program holds weeklong writing retreats on campus. Participants spend the week on campus but away from their office, with lunch provided and a writing coach available. In 2012, women scientists from the Big Ten schools were invited to participate. Additional funding was provided so that women could bring their children, and child care was available on campus. Participants preferring to leave their children at home were offered supplemental funds to cover the extra care costs incurred during their absence.

15.3.2.3. On campus: Weekly. Several of the GAIN writing retreat participants have instituted their own writing retreats. These take many forms, from small groups with regular meeting times to individuals finding a regular place to write away from their office and lab. One academic couple now takes weekend writing retreats, during which they check into a hotel, write in the morning, and explore their surroundings in the afternoon. All of these retreats are similar, however, in their emphasis on eliminating distractions, regularity, and defined periods of time. It's important to note that all of these programs for increasing writing productivity confer benefits not solely on the participants. Equally important is the fact that information about the benefits of regular writing practice is being passed along by these women in their graduate students, the next generation of geoscientists.

15.3.2.4. Online. There are several for-pay writing assistance programs. Three are mentioned here, selected in part on the basis of longevity. These programs may change and therefore differ from what has been described here at the time of publication, but all have been available for over five years.

15.3.2.4.1. The academic ladder provides a variety of individual and group writing coaching services (www.academicladder.com). This includes the Academic Writing Club, which uses research by Boice and others to help participants maximize their output through accountability and individual coaches.

15.3.2.4.2. The national center for faculty development and diversity (http://www.facultydiversity.org) offers a number of different programs. Members (institutional and individual) receive the weekly Monday Motivator, monthly tele-workshops, moderated discussion forums, writing challenges, and facilitated learning communities. The Faculty Success program is considerably more expensive and lasts for 15 weeks. It provides assistance with maximizing productivity and maintaining a work-life balance. They have recently added a Career Center.

15.3.2.4.3. The joyful professor (http://www.joyful-professor.com) is designed to help academics refocus their visions, needs, and goals. It provides worksheet exercises and coaching to allow people to find their "soulful values" and with that focus succeed personally as well as professionally.

15.4. Summary of Helpful Steps

The following are helpful steps that could be supplied by a department, department chair, academic dean, or provost to his or her faculty:
- Help/require a new hire to write at least six hours a week.
- Establish writing retreats on your campus, with child care.
- Establish and provide funds for off-campus writing retreats.
- Inquire about graduate students' or new professors' writing habits. If they have not developed a regular practice, suggest including funds for an online writing program in the startup package. (No formal study of these programs has been published.)
- Provide copies of two or more of the writing resource books listed below to all new faculty.

RECOMMENDED BOOKS ON WRITING AND WRITING PRACTICE

Boice, R. (1990), *Professors as Writers: A Self-Help Guide to Productive Writing*, New Forums Press, Stillwater, OK.
Clark, R. P. (2008), *Writing Tools: 50 Essential Strategies for Every Writer,* Little, Brown and Co., New York.
Silvia, P. L. (2007), *How To Write a Lot: A Practical Guide to Productive Academic Writing*, APA, Washington, DC.
Zinsser, W. (2006), *On Writing Well, 30th Anniversary Edition: The Classic Guide to Writing Non-Fiction*, Harper Perennial, New York.

REFERENCES

Bentley, P. (2011), Gender difference and factors affecting publication productivity among Australian university academics, *J of Sociology*, *48*, 85–103.
Boice, R. (1984), Contingency management in writing and the appearance of creative ideas: Implications for the treatment of writing blocks, *Behavior Research & Therapy*, *21*, 537–543.
Boice, R. (1990), *Professors as Writers: A Self-Help Guide to Productive Writing*, New Forums Press, Stillwater, OK.

Boice, R. (1994), *How Writers Journey to Comfort and Fluency: A Psychological Adventure*, Greenwood Publishing Group, Westport, CT.

Boice, R. (2000), *Advice for New Faculty Members: Nihil Nimus*, Pearson, Upper Saddle River, NJ.

Brande, D. (1934, reissue 1981), *Becoming a writer*, Tarcher, New York.

Clark, R. P. (2008), *Writing Tools: 50 Essential Strategies for Every Writer,* Little, Brown and Co., New York.

Cohen, G., J. Garcia, N. Apfel, and A. Master (2006), Reducing the racial achievement gap: A social-psychological intervention, *Science*, *313*, 1307–1310. doi: 10.1126/science.1128317

Eigenfactor.org, accessed November 20, 2012.

Fox, M. F. (1983), Publication productivity among scientists, *Social Studies of Science*, *13*, 285–305.

Gutierrez y Muhs, G., Y. F. Niemann, C. G. Gonzalez, and A. P. Harris (eds.) (2012), *Presumed Incompetent: The Intersections of Race and Class for Women in Academia*, Utah State University Press, Logan, UT.

Holleran, S., J. Whitehead, T. Schmader, and M. Mehl, (2011), Talking shop and shooting the breeze: Predicting women's job disengagement from workplace conversations, *Social Psychological and Personality Science*, *2*, 65–71.

Hunter, L. A., and E. Leahey (2010), Parenting and research productivity: New evidence and methods, *Social Studies of Science*, *40*, 433–451.

Massachusetts Institute of Technology (MIT) (1999), A study on the status of women faculty in science at MIT (1999), *MIT Faculty Newsletter*, *11*, 4. http://web.mit.edu/fnl/women/women.html

Monroe, K. R., S. Ozyurt, T. Wrigley, and A. Alexander (2008), Gender equality in America: Bad news from the trenches and some possible policy solutions, *Perspectives on Politics*, *6*, 215–234.

Moss-Racusin, C., J. F. Dovidio, V. L. Brescoll, M. J. Graham, and J. Handelsman (2012), Science faculty's subtle gender biases favor male students, *PNAS*, *109*(41), 16474–16479. doi:10.1073/pnas.1211286109

Silvia, P. (2007), *How to Write a Lot: A Practical Guide to Productive Academic Writing*, APA, Washington, DC.

Sonnert, G., and G. Holton (1996), Career patterns of women and men in the sciences, *American Scientist*, *85*, 63–71.

Steele, C. (1992), Race and the schooling of African-Americans, *Atlantic Monthly*, *269*(4), 68–78.

Symonds, M., R. E. Gemmell, N. J. Braisher, L. Tamsin, K. L. Gorringe, and A. M. Elgar (2006), Gender differences in publication output: towards an unbiased metric of research performance, *PLoS ONE*, *1*(1), e127, doi:10.1371/journal.pone.0000127. Accessed October 4, 2012.

Vallian, V. (1977), Learning to work, In S. Ruddick and P. Daniels (eds.), *Working It Out*, Pantheon, New York.

Vedantam, S. (2012). How stereotypes can drive women to quit science, http://www.npr.org/2012/07/12/156664337/stereotype-threat-why-women-quit-science-jobs.   Accessed July 15, 2012.

Walton, G. M., and G. L. Cohen (2007), A question of belonging: Race, social fit, and achievement, *Journal of Personality and Social Psychology*, *92*, 82–96.

Walton, G. M., and G. L. Cohen (2011), A brief social-belonging intervention improves academic and health outcomes of minority students, *Science*, *131*, 1447–1451.

Wilson, R. (2012), Scholarly publishing's gender gap: Women cluster in certain fields, according to a study of millions of journal articles, while men get more credit, http://chronicle.com/article/The-Hard-Numbers-Behind/135236/. Accessed November 4, 2012.

Zuckerman, H., and J. Cole (1975), Women in American science, *Minerva*, *13*, 82–102.

INDEX

Women in the Geosciences: Practical, Positive Practices Toward Parity, Special Publications 70.
First Edition. Edited by Mary Anne Holmes, Suzanne OConnell, and Kuheli Dutt.
© 2015 American Geophysical Union. Published 2015 by John Wiley & Sons, Inc.